Aurora
Model Kits

Thomas Graham

Schiffer Publishing Ltd

4880 Lower Valley Road, Atglen, PA 19310 USA

Disclaimer

An earlier version of this book was published by Kalmbach Books in 1998 under the title *Greenberg's Guide to Aurora Model Kits*.

Aurora is a registered trademark of CineModels, Inc. This book is neither authorized nor approved by CineModels, Inc., Revell-Monogram, Inc. or Playing Mantis, Inc.

Published by Schiffer Publishing Ltd.
4880 Lower Valley Road
Atglen, PA 19310
Phone: (610) 593-1777; Fax: (610) 593-2002
E-mail: Info@schifferbooks.com

For the largest selection of fine reference books on this and related subjects, please visit our web site at
www.schifferbooks.com
We are always looking for people to write books on new and related subjects. If you have an idea for a book please contact us at the above address.

This book may be purchased from the publisher.
Include $3.95 for shipping.
Please try your bookstore first.
You may write for a free catalog.

In Europe, Schiffer books are distributed by
Bushwood Books
6 Marksbury Ave.
Kew Gardens
Surrey TW9 4JF England
Phone: 44 (0) 20 8392-8585; Fax: 44 (0) 20 8392-9876
E-mail: info@bushwoodbooks.co.uk
Free postage in the U.K., Europe; air mail at cost.

Copyright © 2004 by Thomas Graham
Library of Congress Card Number: 2003114663

Designed by Mark David Bowyer
Type set in Seagull Hv BT/Souvenir Lt BT

ISBN: 0-7643-2018-1
Printed in China
1 2 3 4

Contents

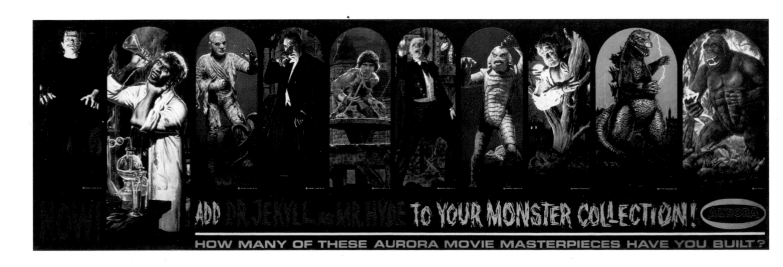

Acknowledgments

For this book I am indebted to a number of individuals. They have made a dozen years of researching and writing a pleasure. I hope that this book is at least partial repayment for their help.

I particularly want to thank the leading families of Aurora for sharing their memories with me: Joseph Giammarino, Aurora's first president; Mrs. Maria Shikes, widow of President Abe Shikes; Stephan Shikes, his son; Mrs. Marie Cuomo, widow of Sales Director John Cuomo; Joanna Cuomo, his daughter; and Charles M. Diker, President of Aurora from 1969-1975.

The ex-Aurora employees and industry associates who contributed to this book are: James Bama, Jack Besser, Derek Brand, John Brodbeck, Anthony Carnaggio, Frank Carver, Robert Chicken, Arthur and James P. Cox, Alvin G. Davis, Blanche (Mrs. Adam) Ehling, Si Friedman, Maurice and David Gherman, Roy Grinnell, Raymond Haines, James Keeler, Henry Kolodkin, Jo Kotula, Mort Künstler, William J. Lemon, Joanne Leynnwood, Mike Meyers, Raymond Meyers, Walter Moe, Richard Palmer, Nat Polk, Anthony Rudisill, Jim Russell, Harry Schaare, Richard Schwarzchild, Donald "Bill" Silverstein, Willem Thomas, and Thomas West. Former Aurora staffer Andrew P. Yanchus published many articles in his series "The Runner" that were of great value.

The men from Revell-Monogram who helped bring the narrative up to the present are: George L. "Jim" Foster, Robert Johnson, William Lastovich, and Ed Sexton. Dean Milano of Revell-Monogram is the custodian of Aurora's history—and, in fact, all plastic modeling. Several items from his collection are pictured in this book.

Tom Lowe, president of Polar Lights, generously shared his thoughts on the operations of his company.

Andrew Eisenberg provided information about his company CineModels, Inc.

Anthony Taylor's regular series "The Assembly Line" in *Toy Shop* magazine is a tremendous source of esoteric information on all phases of plastic modeling.

Fellow kit collectors who shared information and loaned kits are: Bob Bergedick, John Burns, Tom Carter, Mark Carthaarp, Craig Clements, Jim Crane, Bart Cusick, Rick DeMeis, Bob Frongillo, Don Griffin, Mat Irvine, Jerry Irwin, Pat Luther, Pat Mazza, Vince Paulauskis, Kirk Robinson and K. "Buc" Wheat. Tony Ortega's "phrankenstign" web site in the unofficial catalog of Polar Lights model kits.

Finally, I would like to thank my longsuffering wife Susan who has surrendered a lot of closet space to stacks of model kits, and who has patiently listened to her husband's stories about the marvels of plastic model biplanes, sports cars, ships, and monsters.

Chapter 1
In a Garage in Brooklyn

In the fall of 1952 Dwight D. Eisenhower was elected President of the United States, bestowing grandfatherly reassurance on a nation enjoying unprecedented affluence but living in the shadow of the atomic bomb. Veterans from World War II and Korea resumed their lives, moving to the suburban world of ranch style homes with new Chevies, Fords, and Studebakers parked in the car ports. Their kids rode bicycles, shot Daisy air rifles, watched *Sky King* on TV, listened to 45 rpm records, and read comic books. And they made model airplanes. Not old fashioned wooden models, but neat little .69 cent plastic kits that they could buy at the dime store, bring home on their bicycle, and build in a half hour. Then they could play with them until they broke and go buy another, or they could hang them from the ceiling of their room and imagine they were dogfighting the Red Baron or flying in "Mig Alley."

By the 1960s kids had another toy to play with—slot cars. The younger brothers of kids who had received electric trains for Christmas in 1952, got electric slot car sets in 1962. The best slot cars were made by Aurora Plastics Corporation, the same company that made model airplanes, as well as those models of Frankenstein, Dracula, and other movie monsters.

Today's grownups who spent their childhoods in the 1950s, 1960s, and 1970s are rediscovering the hobbies they outgrew years ago. One of the first things they learn is that Aurora Plastics Corporation is no more. Aurora is a name that elicits passions like few others. There is a mystique about Aurora. Perhaps Aurora's model kits were not the most accurate or complicated, but they had style and conveyed a sense of excitement that other companies' models didn't quite elicit. Often damned by serious modelers, Aurora was loved by millions of boys hooked on Aurora kits.

Aurora started selling model airplanes about the time Eisenhower won the presidency. The company grew, diversified its line of hobby products, and became the largest maker of hobby crafts in the world by the 1960s. Then in the 1970s Aurora became part of Nabisco Corporation's empire. Under new ownership Aurora began manufacturing games like "Skittle Bowl" that were big sellers like the slot cars and models. However, in the tumultuous business climate of the seventies, Nabisco decided to break up Aurora. One of the most successful hobby companies of all time disappeared.

Today, if you know what to look for, you can still find some of Aurora's old models boxed under the Glencoe, Polar Lights, or Revell labels. However, almost all of Aurora's once vast assortment of models now belong to the ages. Most will never be produced again. For collectors of Aurora models the excitement has not stopped; it's just that finding an Aurora kit is more difficult than riding your bike to the nearest Woolworth.

The story of Aurora spans the years 1950 to 1977— a brief twenty-seven years. Within that time are compressed many of the major developments in the modern hobby industry. Aurora was a leader in making plastic model kits a staple item in hobby shop inventories. Aurora's success paralleled the rise of the modern hobby industry, and its demise reflects evolutionary changes in the business.

Aurora's rise to prominence is the story of the partnership among three men: Abe Shikes, Joseph Giammarino, and John Cuomo. For the first twenty years of Aurora these partners—collectively "the boys"—were Aurora Plastics Corporation.

Abe Shikes was born in Czarist Russia in 1908. He left shortly after the Bolshevik Revolution to live in Western Europe and then emigrated to the United States. In New York City he picked up an education while working at a variety of jobs. He stood only about 5'4" tall, but made up for his size with boundless ambition to excel. Shikes became an aggressive, chain-smoking, fast-talking businessman with a talent for starting and running companies. "You must have a feeling for good things," he once declared, meaning things that would sell.

He ended up in the costume jewelry business, becoming president of Marvel Jewelry Company in 1939. During World War II he served in the European theater as a humble PFC in the commissary department. After the war he returned to Marvel, but also went into a partnership at Empire Plastics in the Bronx. By this time, his companies were producing plastic beads for necklaces, and his path crossed that of Joseph Giammarino.

Eight years younger than Shikes, Giammarino's personality trended in opposite directions from Shikes. While Shikes was talkative, Giammarino was a private person absorbed with mechanical work. A handsome man with dark, wavy hair, Giammarino had graduated from Brooklyn Polytechnic with a degree in engineering. He stayed at home during the depression years to help in the family jewelry business. As World War II approached and the government began putting restrictions on sales of copper and brass, Giammarino experimented with plastic as a substitute material. During the war he worked as a civilian employee of the Air Force inspecting aircraft parts. In 1950 Shikes approached him with a proposal to start a new business.

The new corporation gained a charter from the state of New York on March 9, 1950 and began operations that August. Giammarino became president and owned one-quarter of the company's stock, while Shikes and two other investors held the rest of the stock as silent partners.

Picking a name for the firm typified the collaboration between Shikes and Giammarino. Shikes said the name must begin with the letter "A" because that would put the company first in the trade directories. Searching for a name, Giammarino remembered that Ben Franklin had pronounced his benediction on the work of the 1787 Constitutional Convention by observing that the sun carved on the back of the president's chair depicted a rising sun. "I wanted my company to be a rising sun," declared Giammarino. Thus he picked *Aurora*, the Roman goddess of the dawn. His pen and ink sketch of the Aurora sunrise with light beams radiating out became the first company logo.

John Cuomo joined Aurora during its second year of operations. Cuomo was older than Shikes and Giammarino, having been born in Italy in 1901. Throughout his life he had been a salesman, starting with *Colliers Magazine*, then selling poster advertising on trolley and subway cars, and finally boosting Iodent Toothpaste. His effervescent personality and natural optimism made him very likable—and a good salesman. He had already been working for Shikes at Empire Plastics.

Cuomo joined Aurora in January of 1952 when it reorganized. The three principles sealed a new agreement over dinner with their wives at a restaurant on Sheepshead Bay in Brooklyn. Cuomo came into the company with a ten percent share of the stock, while Shikes and Giammarino split the remaining ninety percent between them. Shikes replaced Giammarino as company president.

At the beginning Aurora operated as a contract shop to which other companies farmed out their molding. Naturally, the first customer was Empire Plastics, where Shikes continued as a partner, dividing his time between Empire in the Bronx and Aurora in Brooklyn. Giammarino's relatives in the jewelry business also ordered work done, and eventually Aurora had about 100 other clients. Plastic beads in all shapes, sizes, and colors were the meat and potatoes of the production line, but a variety of novel plastic items also came from the molding machines: baby food dishes, plastic banks, cutting boards, and "relief tubes" that airplane pilots took with them for use on long flights.

Aurora seemed to catch a big break when Shikes went to Woolworth and wrangled a huge order for Empire to supply them with small plastic clothes hangers. Aurora would do the manufacturing. However, the clothes hang-

John Cuomo, Joe Giammarino, and Abe Shikes pose for a publicity shot with one of Aurora's first models, the Lockheed F-90 (33 $55-175).

ers turned out to be a one-shot success. Woolworth gave Empire no reorders once customers discovered that clothes slipped right off the hangers!

Then, by happenstance—as Shikes told the story—an outside salesman saw the plastic arm of the clothes hanger without its metal hook and thought it looked like a toy bow. Inspired by the concept, Shikes notched the ends of the hanger, stretched some string between the ends, and stuck a piece of dowel through the hole where the hook went to serve as an arrow. It worked! The bow sold for .29 cents and inspired a smaller bow of the same design that sold for .10 cents. The bow and arrow sets sold seven million copies. Thus—thanks to Shikes' resourcefulness—Empire and Aurora had a huge success after all.

Aurora made its products in a converted garage on 62nd Street in Brooklyn, that, as one worker described it, "looked like it dated back to the origins of the automobile." The neighborhood was far from ideal and visitors to the facility placed their parked cars at risk. However, Aurora thrived in this environment. The garage measured just 40' x 100', but as business expanded, an addition was made to the garage and more space was rented in nearby buildings. Aurora's production came from two medium-size injection molding machines. From the very beginning Giammarino kept the machines running twenty-four hours a day. Too much time was wasted in shutting down and then restarting. With proper planning, the only time a machine needed to be stopped was to change molds or make repairs. The system was geared for mass production. The down side to this philosophy of production was that Shikes, Cuomo, and Giammarino's commitment to the company became an around-the-clock affair.

The Empire "coat hanger" bow and arrow, left, inspired Aurora's Big Chief imitation and the larger Brave Warrior bow.

Another man integral to Aurora's success joined the company in its second month of operation: Henry Kolodkin. He was an experienced operator of injection molding machines, with some training as a tool maker. Kolodkin became Giammarino's right hand man and production supervisor. Every work day began with Giammarino and Kolodkin sitting down over black coffee to outline the day's activities.

In 1952 Aurora took a tentative step toward becoming more than just a job shop for other companies. Shikes left his position at Empire Plastics, and Aurora brought out the first product sold under its own trademark, a .10 cent Big Chief bow and arrow set that closely resembled the smaller set Empire had been selling. The next year Aurora went further and sold a larger, .58 cent Brave Warrior archery set. During the summer of 1953 Aurora began marketing the Aquamatic squirt gun, a toy Giammarino designed. It didn't sell very well, but by that time the company had discovered a far more profitable product: model airplanes.

Model Airplanes

The man who introduced Aurora to the hobby industry was Raymond Haines, the "H" in HMS Associates of Willow Grove, Pennsylvania. HMS designed "patterns," the prototypes of items that a tool company would copy to produce metal production molds. As part of his job Haines solicited business for HMS's craftsmen. In the spring of 1952 he went to the boys at Aurora with an idea.

A heavy man with crew-cut hair, Haines loved everything having to do with aviation and automobiles. He piloted his own plane and drove a Corvette. When he was not flying through the air or along the highways, Haines flew model airplanes. What he had to show Aurora's executives were two plastic scale model airplanes that Hawk Model Company of Chicago was making.

Abe Shikes, left, and Ray Haines look over the prototype of the Douglas DC-8 (386 $90-110).

Haines sat down with Giammarino and Shikes in the little office the two men shared. "I think you guys should really look over a business called the hobby business," he explained. Haines showed them the models and asked at what price they thought they could manufacture them. Giammarino took the plastic parts from one of the kits, weighed them on the scale he kept on his desk, and said he could make it for nine or ten cents. A few calculations showed them that they could clear a .45 cent profit on a $1.00 kit. "There's a business for you if you want it," said Haines. Giammarino replied: "What are we waiting for!" In their haste, the

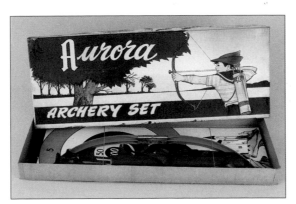

The Archery set was the most elaborate product in Aurora's bow and arrow line.

Aurora men took the shortcut of copying Hawk's two kits. That was, admitted Giammarino, "the only bad thing."

In their rush to enter the hobby field, Aurora made its first two kits, the F9F Panther (22) and the Lockheed F-90 (33), by having a mold company simply copy the parts of Hawk's kits. The only important change Aurora made was to eliminate the separate pilot figures, which Hawk had molded in olive plastic.

Aurora even copied Hawk's instruction sheets and used a similar one-piece folded cardboard box. The only improvement in Aurora's kits was inclusion of a small tube of styrene plastic cement. At the time plastic cement could not be found in most stores, and some hobbyists made the mistake of trying to use the same glue that had worked on wood kits. To meet this need, Aurora began selling its own plastic cement in 1953.

Aurora's kits were in stores by the fall of 1952. At the hobby industry's annual convention, Phil and Richard Mates, Hawk's owners, confronted the Aurora men and asked why they had stolen their kits. (However, when the author talked with eighty-seven year old Richard Mates in 1990, he declined to criticize Aurora, saying only that Aurora offered "good competition.") To protect themselves against being "knocked off" again, the Mates brothers began to hide the word "Hawk," spelled-out in Morse Code, in the rivets of their models and the lines of their instruction sheets.

Aurora's first two model airplane kits (22 $160, 33 $175) came in plain flip-top cardboard boxes that were standard in the hobby industry at the time.

Another key to building a business is distribution. Hawk's kits sold for $1.50 and $2.00 because Hawk had only small volume distribution to hobby shops. However, Revell of California had recently begun selling its Highway Pioneers model cars for .69 and .89 cents. This low cost made them attractive to dime stores and variety stores like Woolworth, W. T. Grant, and Kresge. Aurora already had a relationship with Woolworth, and John Cuomo went to Grant to make a sales agreement. Aurora could match Revell's base price of .69 and .89 cents with a product—model airplanes—that Aurora thought would be more popular than model cars.

Of course, Aurora wanted to sell its kits in hobby shops, too. Haines helped Aurora gain entrance into the hobby industry by introducing them to Mattie Sullivan, his neighbor in Willow Grove. Sullivan, a thin, hawk-faced man, sold flying model accessories and was well known throughout the hobby industry as a hyperactive character. Aurora hired him as merchandising manager and assigned him the task of making the right contacts in the hobby fraternity. "We were greenhorns," admitted Giammarino.

Aurora went to two other men for advice: Nathan and Irwin Polk, operators of Polk's Hobbies, a five-story hobbyist's paradise on New York's Fifth Avenue. They had opened their business in 1932 and were regarded as godfathers of the modern hobby industry. (Their store appears in the Christmas shopping scene in the *Godfather* movie.) The Polks had refused to carry Aurora's two pirated kits, but thereafter were important customers, as well as friends. Shikes talked on the phone regularly with Nat Polk, asking for advice, and, in return, Shikes always made sure that Polks' orders for Aurora products were filled promptly. When the Polks' Manhattan store closed in 1991, an era in hobby history ended.

By the time Aurora was founded, a small, but thriving hobby industry had developed in the United States. Aurora was one of many companies operating out of a garage. Before the 1950s, the basic material used in hobby crafts was wood, but making something out of wood required skill and a workshop full of tools. If a kit manufac-

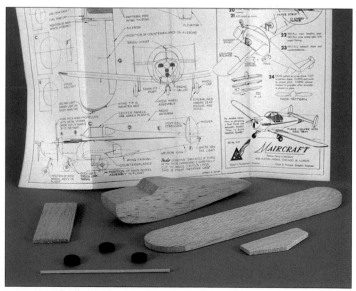

Wooden model kits from the 1930s and 1940s had contained elaborate construction plans, but just a few partly-formed wooden parts and a few rubber, wire, or plastic accessories.

turer tried to make things easier by pre-shaping the wood and adding metal, rubber, or plastic parts, the cost of producing the kits multiplied. Thus model makers were either mature craftsmen or frustrated boys with razor blade cuts on their fingers.

Plastic kits were the answer. Frog of Great Britain introduced the first all plastic models before World War II, and with the end of the war the age of plastic models dawned. The problem with the early plastic kits was high cost, since tooling production molds and buying molding machines quickly added up to a great expense—until Revell and Aurora solved the problem by mass production and mass distribution. Once the mold had been made, its cost could be amortized quickly because it took only a few, semi-skilled workers to produce huge quantities of kits.

In the early 1950s the American hobby industry consisted of a surprisingly close-knit family. Executives from various companies knew each other on a first name basis, and an employee of one company this year might be working for a competitor the next. They all read *Craft, Model and Hobby Industry*, the trade journal. The third week in January each year the manufacturers, distributors, and retailers would gather in Chicago for the convention of the Hobby Industry Association of America (HIAA). Here businessmen and hobbyists would show off their new products for the coming year, scout what the competition was up to, and talk shop.

A few weeks later many of these same people would assemble at 200 Fifth Avenue, New York, the twin Toy Center buildings, for the mammoth Toy Fair. By the time these two conventions were over, the distributors and retailers had decided which products they would order for next Christmas, and the manufacturers knew which items they were going to have to hurry into production and which ones they were going to cancel.

What Aurora had to show in Chicago in January, 1953 was a line of 1/48 scale aircraft. Haines had suggested that Aurora's premier original kit be a stunt biplane. However, Giammarino had his own strong idea what the first plane should be: a Curtiss P-40. He felt sure that the grinning shark's mouth painted on the nose would have irre-

sistible appeal to boys. Also, he said it should be called the "Flying Tiger." Giammarino was right about the Flying Tiger's popularity. The mold stayed locked in the injection machine for nine straight months, pumping out parts. The P-40 (44) would remain in the Aurora catalog down to 1974, and both Giammarino and Kolodkin figured it might be Aurora's all time best selling kit.

Another thing Aurora did with six of its early kits was mold them in bright colors—simply to increase their appeal to young boys. Rather than try to match the colors of real aircraft—silver, olive, blue, white—Aurora gave youngsters a bright red Messerschmidt (55), a green Yak-25 (66), a yellow Japanese Zero (88), a P-38 (99) and a Spitfire (20) in blue, and a Focke-Wulf (30) in black. The effect was startling.

To show young model builders where to place the decals, Aurora molded raised locater outlines on the wings and fuselage. Serious modelers threw up their hands at these departures from true modeling accuracy, but Giammarino felt that they would paint their models anyway, and the raised decal markings could be easily sanded off.

The first release of the Spitfire (20 $110) came in robin's egg blue. Second-issue ME-109s (55 $45-160) were molded in metallic burgundy plastic.

The first issue of the P-40 (44 $25-150) came in gray plastic and lacked landing gear. Subsequent issues came in olive plastic with landing gear, a belly fuel tank, and bombs.

Aurora's kit numbering system simply doubled the digit indicating the order of each kit's release. They skipped 11 because it could be confused with the Roman numeral II; thus the Panther became 22, the F-90 was 33, the P-40 was 44, and so on. When they got to number 99, the P-38 Lightning, they started over again, but put a zero after the number indicating release order: 20 Spitfire, 30 Focke Wulf, 40 Hellcat—up to 80, the SNJ Navy trainer. After that, the numbering system became more complicated.

The HMS workshop was near the Willow Grove Navy air station; so the designers of the SNJ model (80 $25-65) gave it Willow Grove Navy Station unit markings.

One of Aurora's most popular early kits depicted an airplane that never existed. In June of 1953 Aurora released the Yak-25 (66), describing it as one of the Soviet Union's latest and most dangerous fighters. The aircraft also found its way into plastic kits from other companies and even into bubblegum cards—but there was no Yak-25. Aurora subscribed to all the airplane and popular science magazines, as well as *Jane's* military hardware books. These had carried stories of a rumored Russian fighter that would replace the MIG-15. However, the stories proved to be nothing more than rumors. In 1954 Aurora changed the name on its kit box to MIG-19 and continued to sell it until 1970.

Box Art

One of the most important aspects of model sales is the box art, since the unassembled parts are not much to look at. Thus Aurora sought a better package than the conventional gray boxes of the Panther and F-90. Mattie Sullivan had the answer. He took Abe and Joe to the town of Fairton in southern New Jersey to meet George

Burt, who ran a small print shop. Burt, who friends described as an "unlettered genius," worked out of a converted chicken shed. He had become expert at extracting superior results from a basic four-color lithographic printing press. Aurora hired him to do the "wraps" for their boxes. A "wrap" is the printed paper which is folded and glued around a cardboard box. This approach results in a sharper image than printing directly on cardboard.

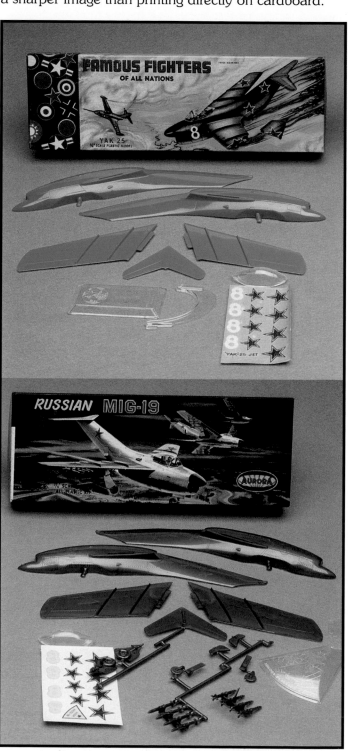

The Yak-25 (66 $70-200) first appeared in basic green plastic with no landing gear. Transformation into the Mig-19 brought metallic green plastic, landing gear, and rockets. Jim Cox's original box art gave way to Jo Kotula's dramatic twilight scene.

Burt also printed Aurora's instruction sheets, catalogs, and advertising fliers. Although Aurora became Burt's largest customer, he also did printing for a number of other companies, including Lindberg, Pyro, Renwal, Bachman, Lifelike, and Hubley. Burt would continue his relationship right down to the demise of Aurora in 1977.

Acme Paper Box Company of Philadelphia was the other partner in Aurora's kit box production. Burt would ship his wraps to Acme, where they would be held in inventory. Each afternoon Kolodkin would telephone Acme with an order for the next day's boxes. Acme's trucks would be at Aurora's loading dock at eight the next morning ready to unload. Twelve hours later these same boxes would be filled with Aurora models and on the loading dock again, ready to be shipped out to customers.

The other big idea of Sullivan's was adoption of the trademark "Famous Fighters of All Nations." It appeared on the top of Aurora's boxes, while the rising sun logo was relegated to the box ends. However, as years went by the name Aurora became more prominent and the Famous Fighters name was downgraded. When Aurora adopted a new oval trademark (designed by Chick Tatum, maker of Aurora's decals) in the summer of 1957, Famous Fighters was relegated to the logo's borders.

Aurora's most prolific artist was a man whose signature didn't appear on any of his work, but millions of kit builders have seen self-portraits of his hands on Aurora instruction sheets. James P. Cox worked for Burt's printing company from the early fifties through 1972, doing almost all Aurora's instruction sheets, as well as most of the early box art.

Cox had left the hectic life of a Philadelphia commercial artist for the peace and quiet of Burt's rural print shop. The switch to doing assembly instructions for model airplanes was not an entirely new departure for him because he had done "exploded views" of aircraft for the Navy during World War II. For each new kit he would complete a line drawing of the model parts and then overlay it with a transparency containing the shading that gives Aurora's instruction sheets a sense of depth.

The preliminary text for "every single instruction sheet" was done by Giammarino himself, with the help of his son Eddie and Shikes' son Stephan, who served as guinea pigs for testing new kit assembly. After consulting with Giammarino, Cox would complete the instruction sheet. "As I did the instructions, I imagined a kid and his dad building the model on the floor."

Beginning with the P-40 Flying Tiger, Cox did most of Aurora's box art. Two of his paintings are probably the most famous box illustrations in all modeling history: the red airplane against a yellow sky in the Fokker triplane kit (104) and the burning green Fokker D-7 (106) with its pilot screaming in agony. While Aurora periodically updated the box art on its kits, Cox's two Fokker wraps stayed in place until 1976 when Aurora switched to photographs on its boxes.

Cox did a number of other tasks for Aurora, including layouts of the annual catalogs until about 1960. By the late fifties Aurora began bringing in other artists to paint the box wraps, so Cox's work was limited to the instruction sheets. However, he also did art work for other hobby companies when he had the opportunity, and, after his retirement, Cox came back one last time to draw instruction sheets for Addar.

Jim Cox's Fokker D-7 (106 $15-40) box art appeared in hobby shops for nearly two decades, making it perhaps the most famous model illustration of all time.

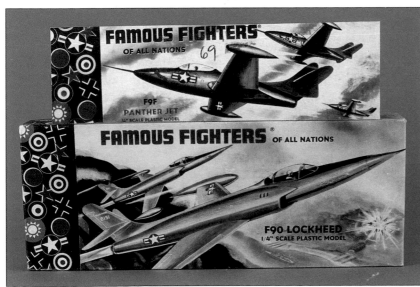

Aurora led the way in dramatic box art. The Panther Jet and F-90 (22 $25-160, 33 $55-175) quickly shed their dull gray first-issue packaging for colorful Jim Cox box art.

Chapter 2
Build Them All!

By the summer of 1953 Aurora started running out of space in its Brooklyn garage. As Henry Kolodkin explained, Aurora had crammed in so much equipment "you just couldn't move." And still the demand for models increased. "We just couldn't make enough. We'd put a mold in the machine, and the orders were coming in so fast that we just didn't stop production. It ran continuously. Every time we got a new item we had to buy another machine. We had to buy another larger machine for the F-90." Some molding projects were farmed out to contract shops.

In addition, Aurora was running-up monthly water bills of $4,000 to pay for the water used to cool the machines. Giammarino hired a well driller to suggest a new plant site. One day an employee passed a building under construction in West Hempstead, Long Island and brought a yellow brick back to show Shikes and Giammarino. The location was perfect. Before the building's floor was poured, Giammarino, Kolodkin, and a few workers went in on a Sunday and dug trenches for the pipes that would run water to the mold machines.

In December, 1953 the move to West Hempstead was made. The new plant had 25,000 square feet and housed eight mold machines. Its only drawback was the address: 44 Cherry Valley Avenue. Shikes thought "Road" sounded better, so Aurora always listed its address as 44 Cherry Valley Road. Despite its rustic-sounding address, the plant stood near the busy Hempstead Turnpike.

Just before the move, Aurora began making some important changes and additions to its line of kits. Giammarino had laid down a rule that ten percent of every year's gross receipts must be reinvested in new tooling. The most dramatic addition to Aurora's line was ship models. "We were jealous of Revell," explained Giammarino. Aurora had developed a growing following among modelers, and to keep brand conscious customers happy, Aurora needed to offer them more than just aircraft.

The Aurora Line

The first ship model was the new atomic submarine *Nautilus* (500), which Aurora labeled a "symbolic model" to reassure a nervous public that it was not giving away secrets to the Russians. The world's first nuclear submarine *Nautilus* became a runaway best seller for Aurora from its first offering in 1953. The kit had very few pieces and was faithful to the real submarine only in a general way. It was based entirely on an article that appeared in the December 1952 issue of *Colliers* magazine. To spice-up the first issue of the kit, Giammarino included his favorite decal from the Flying Tiger airplane kit—the eyes and toothy shark's mouth.

John Cuomo, Joe Giammarino and, Abe Shikes stand in the doorway of the new plant at 44 Cherry Valley Road. The original sunrise logo appeared on the end panels of early kit boxes.

The Nautilus (708 $30-80) had great appeal to kids, who didn't care whether it was an accurate depiction of the real submarine or not.

In 1957 Aurora cut a duplicate mold of the *Nautilus*, added a deck hanger and missile, and issued it as the *Sea Wolf*, America's second nuclear sub the *Sea Wolf* (706). (The real *Seawolf* was totally different from the *Nautilus* and never carried missiles.) The *Sea Wolf* remained in the catalog until 1972, and the *Nautilus* until Aurora was dissolved. As late as 1973 the *Nautilus* had outsold all but a few of Aurora's kits. In that year it outsold the P-40 Flying Tiger 40,000 to 17,000. Considering its long lifetime, the *Nautilus/Sea Wolf* may be Aurora's all time best selling kit.

The destroyer USS *Halford* (480) and the pirate ship *Black Falcon* (210) quickly followed the first release of the *Nautilus*, with the Viking Ship (320) and Chinese Junk (430) right behind. Each of these models possessed toy like qualities that would make them anachronisms in the modeling world within a few years, but every one of them was a great success with the public. Only the *Halford* failed to maintain a place on Aurora's order lists through the 1960s.

Jim Cox's painting for the USS *Halford* (480) captures the toy-like quality of the model.

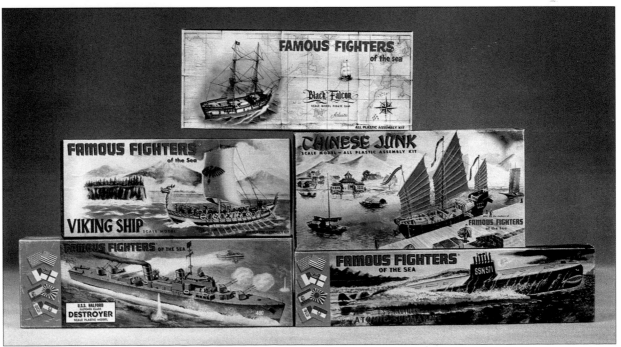

The first five ship models were introduced under the label "Famous Fighters of the Sea."

Low priced kits had made plastic modeling a popular hobby, but by 1954 hobby shop owners began asking for some higher priced items with larger profit margins. Aurora responded with the Martin B-26 bomber (371). With a wing span of 19 inches, it was the largest aircraft kit yet produced by any company and carried a price of $2.59. The advertisements declared that the model was so big "you can even see the rivets."

The Black Falcon, Chinese Junk, and Viking Ship (210, 430, 320) were simple models, but younger modelers just loved them.

Actually the B-26 had no more detail than the rest of Aurora's kits. The oversized rivets were added to create the illusion of detail. During 1954, Aurora sent most of its early kits back to the tool maker to have rivets added. Those kits that had been made without landing gear were given wheels and bombs or rockets under their wings. The final touch to this upgrading was a switch to metallic plastic in some of the planes. The ME-109, which had been striking in red, became resplendent in metallic burgundy. The MIG-19, Spitfire, Panther, Sabre Jet, F-90, and P-38 got the same treatment.

Every single one of these early kits was a tremendous seller. On the back side of each kit's instruction sheet were pictures of every kit in the Aurora "line"—an irresistible inducement for kids to buy all of them to complete their personal collections. In its critical early years, Aurora's men picked the right kits and avoided calamitous investments in expensive tooling for products that didn't sell. The reward for their shrewd hunches was a leading place in the emerging plastic model kit business.

The release of more and more plastic model kits by Aurora and other companies came as a surprise to many old-time hobbyists, who had seen plastic as just a fad. Plastic kits were dismissed as simply unassembled toys, not something that would interest a true hobbyist. Anthony Koveleski of Hudson Miniatures complained that plastic model cars were keeping young hobbyists from learning the skills to build his wooden Old Timers. (Although Koveleski himself was bringing out a line of all plastic Lil' Old Timers even as he complained!)

Bernie Paul, one of the country's pioneer hobby product distributors, lamented that the explosion of plastic model kits had created chaos among wholesalers and retailers. He felt there were too many plastic model kits on the market. However, most hobby industry leaders strongly disagreed, asserting that plastic kits had finally given hobby dealers a product with appeal to mainstream America. Plastic models brought hobby shops from side street hole-in-the-wall operations to the visible prosperity of Main Street. Kits were being sold everywhere: drug stores, stationary shops, bicycle shops—even gas stations.

America's youngsters fell in love with plastic kits. A 1957 survey by *Boys' Life* magazine showed that building plastic models was the favorite hobby of 80% of its readers, far ahead of stamp collecting, which stood in second place. In the 1940s, a similar survey had shown only about 16% of boys built model airplanes. During the Fifties boys assembled models with their friends, showed them off at club meetings, and decorated their rooms with them. The new plastic models were so easy to build and resulted in such nice looking display models that boys could accumulate a whole collection of airplanes, cars, or ships.

The streamline Martin B-26 (371 $45-100) had tremendous appeal to boys who wanted a really big model airplane.

From the manufacturers' point of view, the popularity of models represented a double-edged sword: increased sales boosted company profits, but also encouraged more competition to enter the marketplace. In 1950 only two companies, O-Lin and Hawk, marketed all-plastic model kits. By 1955 twelve larger companies and forty small firms were making plastic kits. Like most emerging industries, hobby plastics was going through a shake-out period of intense competition.

Aurora's competitive strategy was simple: make a quality product and sell it in large volume at the lowest possible cost. Aurora's distinguishing tactic was to be first in the market with a model of a new airplane or ship. Whenever the Air Force rolled-out a new warplane or the Navy launched a new ship, Aurora had a model of it on store shelves almost immediately. Of course, this hurry-up approach led to a good deal of inaccuracy in Aurora's offerings, but the executives at Aurora felt that most boys didn't care about precision.

Revell, the industry leader in sales, took much the same view as Aurora on volume, but stressed detail and accuracy rather than getting their kits on the shelves first. Thus Revell's kits appeared in hobby shops after Aurora's and were priced a little bit higher. Jack Besser, president of Monogram, a smaller, more conservative company, said simply, "We don't want to be the biggest, we want to be the best." Lesser companies tried to find some niche left unoccupied by the major rivals. It was not at all clear which companies would emerge as survivors of the furious competition.

Much of the turmoil among rival companies stemmed from the problem of "duplication"—too many kits of the same subject. Revell and Monogram wanted a P-40 in their company lines, too. This issue always appeared on the agenda at the HIAA conventions. As manufacturers smilingly showed each other the kits that they would bring out during the year, inwardly they calculated their strategies for dealing with the competition. Eventually there evolved a rough gentleman's agreement among the various companies to avoid duplication.

As Ray Haines explained it, "I used to get together with the directors of development of the other companies at the hobby shows. We used to sit around the hotel lounge and have coffee and discuss our plans. Which didn't hurt anything because tooling is very expensive. What we would usually try to do was to feel out some of the other guys and find out what they were making and steer clear, or we'd find out what size they making it in and make it in a different size. If you came head-to-head with the same product in the same size, you were sort of watering down each other's profits." He added, however: "We used to outsmart each other once in a while by lying a little bit, but that was all right."

Evidence of this process can be seen at Aurora in 1953-1954. Announced models of the new F-84 Thunderstreak and F-89 Scorpion never materialized, and Aurora's F-94 Starfire came out scaled at 1/82 rather than the company's standard 1/48 size. Why did Aurora change its plans? Because Revell unveiled its own models of these planes in 1953-54.

Enterprise Models of Mineola, New York, copied Aurora's Yak-25 in red plastic.

The Boeing B-17 (491 $50-60) and the rest of Aurora's line of tiny .39 and .49 cent kits gave youngsters models that were easy to build and fun to play with.

Aurora's little F-94 model (390) launched a series of .39 cent aircraft kits, and it came paired with an F-100 Super Sabre (490) that began a .49 cent series. These simple little kits gave Aurora something to sell at the lowest end of the price spectrum. Because the models were designed to fit into a standard small box, the planes came in a hodgepodge of scales.

In 1957 Aurora began selling its lowest cost kits ever, a .29 cent fighters collection made up of the F-102, F-104, F4D, and F9F-6 (290-293). In 1959 two more fighters joined the 290 series: the F-101 and F-107 (294, 295). The .29, .39, and .49 cent kits were good selling little items, and all of them stayed in the catalog through the sixties.

Hobby Innovations

Aurora remained always on the lookout for novel model subjects, and in February, 1954 an opportunity turned up when a photographer driving by the Convair plant near San Diego spotted an aircraft with its twin propellers pointed straight up into the sky. Stories appeared in newspapers and magazines about this mystery plane, and in March the Navy released photographs of two vertical takeoff aircraft, one being developed by Convair and the other by Lockheed.

Ray Haines and the prototype builders at HMS went straight to work on the Lockheed plane just on the basis of the magazine photographs. Work on the Convair model took a little longer because Convair agreed to supply Aurora with three-view drawings of the aircraft. Within a few months Aurora had both kits on the market. Lindberg also brought out a model of the Convair "Pogo," but Revell did not make either plane. Revell president Lew Glaser bumped into Giammarino at a hobby convention and explained, "When the VTO came out, I thought about making it, but I knew you'd make it and you'd beat us; so we decided not to make it." Aurora's reputation for being first into the market with a hot item had already been established.

The penalty for quick work was usually inaccuracy, but in the case of the Lockheed VTO, Aurora was called into question for being too accurate. Soon after the kit's release, Shikes got a phone call from Washington asking where Aurora got plans for the VTO. Shikes turned to Giammarino with the question, and he simply showed him the magazine photographs. Aurora had scaled the plane from the size of the pilot—and had come in right on the money.

Although plastic kits had proven to be a dramatic success for Aurora, the boys at West Hempstead knew that model planes, ships, and cars had a very distinct limitation in the hobby market: they were purchased almost exclusively by boys. The female half of the population—a huge potential market—showed no interest in the models the boys went crazy over. But women had shown a

Vertical risers were much in the news during the mid-1950s, and Aurora's Ryan X-13 (376 $140-160) exploited the public's interest in these odd aircraft.

great deal of interest in two other hobby fads of the early Fifties: paint-by-numbers and copper etching sets.

Mattie Sullivan thought Aurora might be able to cash in on the female side of the market. He sat down with Ray Haines and the boys at Aurora and gave them a craft lesson. Taking the foil from a cigarette pack, he placed it over a penny, rubbed it with a pencil, and raised Lincoln's image on the foil. Why not do something like that in copper? The boys liked the idea so much that they showed "Coppersmith" right away at a regional hobby show in Baltimore in the spring of 1955. Aurora furnished dealers with 3-D glasses to view their promo display. When the distributors lined up to place orders, Aurora swung into action.

HMS artist Art Fowler drew designs for an extensive line of kits, and sculptor William J. Lemon engraved the images into plastic, being careful not to make the impressions too deep lest the copper split as it was worked into the recesses. After the production molds were made, Giammarino and Kolodkin worked out snags in the production process. The thick plastic squares that held the designs tended to warp as they cooled after being pulled from the molding machines. Soon huge rolls of copper were arriving on the loading dock, ready to be cut into squares ranging from 4" x 4" to 11" x 22". The plastic plates and copper sheets went into boxes along with a wooden stylus and antiquing paint.

Coppersmith didn't turn out to be the smash hit Aurora had hoped for, although it did sell fairly well. In fact, when Aurora tried to drop the line after a few years, hobby shops demanded its return. So it came back in the 1960s as Coppercraft and then again in the 1970s under the original Coppersmith name.

In 1955 Aurora released another item aimed at the craft market: the Shadow Box Aquarium (981), a small framed container with three plastic fish and plastic seaweed. This novel model kit had been produced by Gowland, the design company noted for creating Revell's Pioneers Highway and ship-in-a-bottle kits. The designer of the Aquarium kit was English-born Derek Brand. At the time he was also working on a project to create an electric HO scale slot car set. Brand's Aquarium turned out to be Aurora's first total flop on the retail market, but a few years later Brand perfected his electric slot car system—and that would revolutionize the course of Aurora's history.

Coppersmith made its debut in the 1950s, came back in the 1960s as Copper Craft, reappeared as Tru-Craft Coppersmith in the late 1960s when K & B took over production, and then finally as Safe-N-Sane Coppersmith by K & B in the 1970s.

Aurora made one more stab at the craft market in 1959 when it released Plastik Basketry, kits that replaced the wood and fiber of regular basket weaving with plastic bases and plastic reeds. There was a whole collection of shapes, sizes, and colors in "split-proof, crack-proof, and peel proof" plastic. Unfortunately, the concept did not catch on and dropped out of the catalog three years later.

However, plastic model kits remained the focus of Aurora's energies. By the mid-1950s Ray Haines had been hired away from HMS, but he still spent half his time in Willow Grove working with the HMS staff. He would bring in ideas and model prototypes that had been developed at HMS to the West Hempstead plant, and these would be discussed over lunch. Someone would be sent out to the deli for sandwiches and Cokes, and whoever happened to be available in the plant that day would sit in, eat, and toss around ideas.

The kits that emerged from these bull sessions were the ones that went into production—"with not one iota of market research." Shikes was concerned only with costs of materials and production compared to selling price. If Giammarino and Haines were in agreement, Shikes would say "make six," and a new line of kits would be created. "They didn't fool around," noted Haines.

Once the executives gave the go-ahead on a model, Haines would put the staff at HMS into motion. First, information would be gathered, usually in the form of photographs and line drawings, seldom official blueprints. Draftsmen would make scale outline sketches and then produce more refined engineer's three-view drawings that broke the model down into its component parts.

HMS created the prototype of Henry Hudson's ship the *Half Moon*, but Aurora never developed it as a kit.

The Basket Weaving Kits were a modestly successful attempt to reach the female market.

When these preliminaries were completed, the work was turned over to the team of pattern makers who would divide up the various portions of the model among themselves. HMS's craftsmen carved patterns in acetate plastic, rather than wood as most companies did, because it was easier to work with.

While most companies made their patterns two or three times the size intended for the final model, HMS worked on the same scale as the model. The primary reason for working in a larger size was to make sculpting small details easier, but Haines felt that this was not necessary. However, really fine detail could not be sculpted into a 1/1 pattern. This made little difference in the early days of model making, but would become an important limitation in the 1960s when the industry matured. However, the fitting together of parts and precise assembly were things that Giammarino insisted upon.

When completed, the pattern parts were lightly welded together, and Haines would take the prototype to West Hempstead for final approval. If the model passed this test, photographs would be made of it for use by the box art and catalog artists; then it would be sent on to the mold tooling company. HMS charged Aurora about $8,000 for a simple pattern, like a small sports car, and up to $15,000 for a large, complex kit like a sailing ship.

Molds for most of Aurora's kits were cut by Ace Tool and Manufacturing of Newark, New Jersey. At Ace, the patterns would be disassembled, and for the first time the hollow inside or "core" of the model received attention. Locater pins and holes would be added at this point. The mold shape was sculpted into very hard steel by even harder cutting and polishing equipment.

Ace would be working on several Aurora molds at any given time, and the usual timetable for producing a mold was four or five months. However, if Aurora needed a mold in a hurry, Ace could rush a mold out in as little as three months. The most expensive mold Ace ever did for Aurora, the nuclear powered carrier *Enterprise* (720), cost $71,860. Simple mold repairs and revisions could be made at West Hempstead by Aurora's in-house machinists, but major revisions required sending the mold back to Ace.

Building Better Models

In the summer of 1955, Aurora began releasing a series of classic biplanes that would become the cornerstone of Aurora's aircraft line. The first three biplanes put on the production path were from the "Golden Age" of aviation between the world wars. Although these planes were unknown to most boys in the 1950s, Aurora made them largely because Ray Haines remembered them as exciting aircraft from his youth. The first kit released was the Curtis P6E Hawk (116), followed by the SBC-3 Helldiver (117), and the Boeing P-26 Peashooter (115). They were the most detailed kits Aurora had produced up to that time and had neat features like moving parts.

Aurora, like other hobby companies, sponsored contests designed to promote sales of their kits. The Curtiss P6E Hawk (116 $35-45).

This Chris Craft cabin cruiser would have made a striking model, but Aurora decided not to produce it since pleasure boats don't excite young modelers.

In 1958, these Golden Age planes were joined by three more kits: the M-2 Mailplane (111), Boeing P-12E (121), and the Boeing F4B-4 (122). These kits lacked the operating features of the first three models, but they did include a neat new element: a base part with wheel chocks and a mechanic figure.

These accessories were also standard on the line of 1/48 scale World War I planes Aurora introduced in the mid-fifties. The 1950s witnessed a revival of interest in that conflict's "knights of the air." Articles appeared in men's magazines, old movies like *Dawn Patrol* showed up regularly on weekend television, and Jack Hunter's novel *The Blue Max* became a best seller.

Aurora's Boeing F4B-4 (122 $25-35) was copied by Marusan of Japan, and then this pirated mold was used by Fuji and Entex.

Unfortunately for HMS's model designers, solid information on World War I airplanes was very scarce, and they relied on the few secondary sources available. This resulted in a good bit of inaccuracy. Three-view drawings by William Wylam in old issues of *Model Airplane News* were one source. *Aircraft of the 1914-1918 War*, which contained three-view drawings of the major combat planes, also became a standard resource. Two of Aurora's models, the DeHavilland DH-10 (125) and the Fokker D-8 (135), even carry the serial numbers of planes pictured in this book.

the fans of these models was the singer Mel Torme, an active member of World War I aviation societies. He would telephone West Hempstead and suggest which planes should become models and would go to Polks' Model Craft Hobbies and leave with an armload of Aurora kits.

The Fokker Dr-1 Triplane and Nieuport 11 (105 $15-40, 101 $15-40) were part of a set of six models that launched Aurora's extremely successful collection of 1/48 scale World War I aircraft.

Aurora used *Aircraft of the 1914-1918 War* as its primary resource in designing the Fokker E-III (134 $25-40).

The first set of six World War I aircraft to be done were well known: the French Nieuport 11, the British Sopwith Camel and SE-5, and three German planes, the Albatros D-3, Fokker Dr-1 Triplane, and Fokker D-7 (101-106). These kits flew off store shelves and were destined to occupy page one in the Aurora catalog for years. Among

The World War I series kits introduced two innovations: One was the inclusion of a "ground" base with wheel chocks and a mechanic to created a mini-diorama. The other was molding some parts in black plastic: engines, wheels, machine guns, and propellers. The black parts reduced the amount of painting that young modelers had to do. However, from two to five separate kits shared an auxiliary mold for the black parts, and thus were linked like Siamese twins. A hot-selling kit like the Spad (107) could not be produced in larger quantities than its less popular mold mates.

The famous Spad XIII highlighted the second set of six kits released in 1957. Joining it were the Nieuport 28, German Pfalz D-3, British DH-4, Bristol F2B, and the American Curtiss Jenny (107-114). The sleek Spad fighter, in the colors of the American service, proved to be the most popular of the lot and the rarest kit to find today.

In 1958, Aurora added two twin engine bombers, the German Gotha (126) that had bombed London by night and the British DH-10. The Gotha was a famous and rakish aircraft, and thus a natural for a model. However, the DH-10 was an obscure craft that never saw action in battle. Why did Aurora make it? Simply because it was the only British bomber they could find plans for (from *Aircraft of the 1914-1918 War*), and that was small enough to do in 1/48 scale.

In Great Britain, Playcraft sold the Series B kits in plastic bags with header cards, a packaging style pioneered in England by Airfix.

The German Fokker D-8 (135) and Halberstadt CL-II (136) appeared in 1960. Then, after a pause of four years, Aurora concluded it World War I collection in 1964 at twenty models with the addition of four more kits. The Sopwith Triplane (100) was molded in black to represent the Royal Navy's "Black Fight" aircraft—although the real planes had only black trim. The Fokker E-3 monoplane (134) was made larger than the standard 1/48 scale of the series because a true scale model would have been too small for the standard box Aurora used for the other kits. The French Breguet 14 bomber (141) and German Albatross C-3 (142) completed the line. The Albatross would be the last new 1/48 scale model from the First World War to be issued by any company until the 1980s when interest in vintage aircraft revived.

Aurora photographed the prototype of the Breguet 14 (141 $40-50) for use by the art and advertising departments.

The Patton Tank (301 $25-50) sent Aurora's series of military kits off to an excellent start.

For some reason, model companies avoided making models of army tanks and other equipment until after the middle of the decade. Then, in 1956 Aurora issued some of the earliest military vehicles on the market: three Korean War/World War II era tanks, the M-46 Patton, German Panther, and Russian Stalin (301, 302, 303). These kits, and those that followed, are among the best detailed kits Aurora ever produced. In fact, the Patton represented the most expensive model Aurora had tooled up to that time, with a bill from Ace totaling $27,200. The designers at HMS gave the tanks moving parts and flexible vinyl treads.

Following the release of its first three tanks, there was a pause, but then in 1959 Aurora issued the 8" Howitzer, Long Tom Cannon, M8 Munitions Carrier, and M8 with 8" Howitzer (307, 308, 309, 310). The next year it brought out the British Centurion tank (300) and the LaCrosse missile on the Munitions Carrier (311).

HMS made this pattern for an Army half-track, but Aurora never turned it into a kit—perhaps because Monogram brought out its own half-track.

Other companies were also releasing armor models at this time, but each company selected its own scale. Aurora went with 1/48 scale to match its aircraft, but most others designed theirs to larger scale standards. This meant that model builders could not cross company lines to assemble a unified collection of military vehicles as they could, for example, with 1/48 scale aircraft. Perhaps partly for this reason, interest in military vehicles faded. In 1961 Aurora took all its military vehicles out of the catalog, except for the three original tanks and the Centurion. This was a loss for military modelers because HMS had made patterns for a half-track mounting an anti-aircraft battery and an Honest John rocket on a mobile launcher that never made it into production.

Hitting Stride in the Market

In the final years of the 1950s, the plastic modeling boom entered a new phase. Because of cutthroat competition, product proliferation, and kit duplication, companies had to rethink their objectives. Monogram president Jack Besser decided that there were too many kits on the market and that henceforth Monogram would make fewer kits, but better quality kits. Besser also looked at the potential for sales to adults. In analyzing sales of his very popular 1955 Cadillac kit, he found that seventy percent of the kits had been bought by adults. Besser suspected that other kits being bought by adults "for the kids" were actually taken home and built by grownups.

Revell's Lew Glaser disagreed with the idea that the kit market was glutted. "Unquestionably," he declared, "despite the enormous expansion that has already taken place, the market for authentic models has barely been scratched." Revell would continue its high volume sales strategy, and, indeed, in 1955 promised to bring out "a kit a week." In seeking new customers, Revell also looked to older folks. Its "Old Ironsides" ship model was one kit designed with adults in mind.

Aurora always behaved a bit differently from other companies and followed its own course. The management at Aurora remained confident that it could out-produce and undersell anybody. The question was: who would buy the products? The general wisdom within the industry held that the kit market centered around 11 to 14 year olds, but Aurora's sources in retail stores told them that kit buyers were younger.

Aurora had started out with simple kits aimed at youngsters, and nothing had happened to change their minds. Joe Giammarino said, "Our kits are made for kids from six to ninety-six"—but the emphasis focused on six. Giammarino wanted to keep things simple, and this philosophy was reflected in Aurora's kits. Haines knew what was wanted and saw that the pattern makers at HMS followed the rule of "not too much detail."

As Giammarino saw it, there was a practical trade off between the amount of detail and kit prices. If you wanted to keep prices low, you couldn't have complicated molds. "We had what we used to call the screwball hobbyist," he explained. "You could never give him enough detail." Or as production manager Frank Carver explained, only the purists were interested in absolute accuracy. "You have

to be a nut to pick up a caliper or micrometer to measure accuracy. We could not build a company of the magnitude we were trying to achieve by trying to appease the few rather than the many."

Aurora was set on producing a low-cost, simple, well-made kit, for ten year olds. It would make them the largest volume model company in the world within a few years—and it would also drive serious modelers who were nuts for accuracy into fits at the mere mention of the name Aurora. (Actually, because Aurora left no stone unturned in the search for sales, it would soon make some kits, like its large-scale ship and car kits, for adult modelers, but even these were simpler than their Revell counterparts.)

Aurora Overseas

Another way Aurora expanded sales was by opening foreign markets. It entered the European field through the initiative of Franz van Breemen, a pilot for KLM airline, who, beginning about 1954, would pick up kits in New York and carry them back to Holland along with the passengers on his DC-6. In Amsterdam, Aurora's models were sold by Willem Thomas, operator of Hobby Holland, a shop and distributorship.

Among the first kits imported into Holland were the German Messerschmitt and Focke Wulf. When Thomas displayed these planes, with the Nazi swastika on their tail fins, some of the local Dutch were outraged. Police were called out to guard the hobby shop, and Thomas held a press conference to explain that these kits had no political significance. Of course, Thomas, the clever salesman, was quietly pleased with the "free advertising."

The Focke Wulf FW-190 (30 $30-110) first appeared with a swastika on its tail fin, but later issues substituted a cross for the swastika.

Meanwhile, back in the United States, Aurora received a request from the State Department to remove the swastikas from their models. Giammarino grumbled that the insignia were a part of history, but he agreed to comply, and the decals were changed to crosses. For a while Aurora still had some of the old stock on hand and distributors could order "swastika" or "non-swastika" kits. The removal of the Nazi emblems also solved the problem of exporting kits to other European countries that had outlawed swastikas.

In 1957 Aurora decided to strengthen its position in Europe by setting up a subsidiary corporation. Shikes, Giammarino, and Cuomo visited Holland to settle things with van Breemen and Thomas. The new corporation, Aurora Plastics Netherlands, was two-thirds owned by the parent company. In the summer of 1958 Aurora Netherlands opened for business in a simple shed building with a quaint metal smokestack on the roof. Located in a meadow near Nijkerk, the plant had plenty of room to expand—which it quickly did as sales boomed.

During its early years of operation, Aurora Holland imported all its kits, packaged in plastic bags, from the Unites States. Nijkerk had a four color printing press and box making machine for printing the box art and packaging the kits. Six language instruction sheets and the annual European catalogs were also printed on their press. Color art plates for the kit boxes and catalogs were done by George Burt's company in the United States and the plates sent to Holland.

After a few years, Aurora began to send some of its model molds over to Europe, as well as bagged kits. Since Holland allowed tooling to enter the country tax free, it was less expensive in some cases for Aurora to export the molds than the kits. Aurora's molding in Europe was done by contract mold shops, first in Holland and later in Austria, where it could be done more cheaply. After the molds had produced a run of kits, they would be returned to the United States. Aurora Netherlands' kits could be sold in the Common Market duty-free because more than half their value had been created in Europe.

In Great Britain Aurora's kits were distributed by Mettoy, the maker of Corgi diecast metal cars and other products sold under the Playcraft name. Most of the Aurora kits sold in Great Britain were manufactured in the United States, but some were made in England from molds shipped there. Kits made in England are marked on the boxes "Made in Great Britain," and sometimes on the plastic of the kits themselves. The most interesting models done by Playcraft were the Series B kits of the first nine World War I aircraft. These were not boxed, but were bagged, with the instruction sheet folded over and perforated to form a header card for hanging the kit on a display rack.

In the 1950s many forecasters speculated that someday Dad would fly off to work in his personal aircraft. Helicopters for Industry captured this vision in the world's first copter kit, the Hiller Hornet (501 $130-150).

Acquisitions

While Aurora, Revell, Monogram, and some other companies were finding ways to prosper in the hobby business, other companies decided to call it quits. Aurora benefited from this attrition by stepping in to buy the molds of companies exiting from the field. By the early 1960s Aurora had absorbed tooling from seven companies.

Playcraft marketed Corgi diecasts, Aurora plastic models, and introduced the "Highways" slot car system. Highways didn't catch on in Europe, but in the United States Aurora turned it into "Model Motoring"—America's most successful hobby product of the 1960s and 1970s.

One of the earliest sets of molds Aurora acquired came from a company located just a few miles down the turnpike in Hempstead. Dr. Eugene Rodin, a physician, had issued the first helicopter models done by any company back in the summer of 1952, before Aurora made its first kits. He called his firm Helicopters for Industry and operated it as a hobby sideline while he continued his practice.

The first kit his company brought out was a model of the tiny Hiller Hornet, and it sold for $2.98, a price typical of the early days of plastics. Eventually Rodin expanded his line to include five whirlybirds, but stretched the number of kits by molding them in Air Force, Navy, Marine, Army, and civilian colors. In 1955 he approached Aurora and sold them his tooling. Giammarino, believing that kids are fascinated with names, put the helicopters into the Aurora catalog and added nicknames: "Ram Jet, Army Mule, Windmill, Work Horse, and Egg Beater." These kits would remain in the catalog until the mid-1960s, when they were phased out and new helicopter models replaced them.

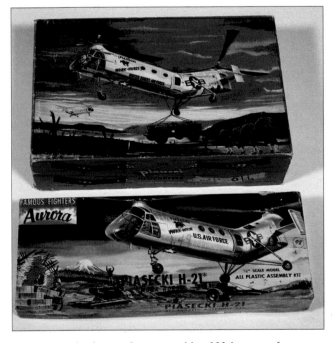

Aurora bought the production molds of Helicopters for Industry and made copters like the Piasecki H-21 (504 $80-100) a cornerstone of its model kit offerings.

Until 1957 Aurora avoided the automotive realm, leaving that field to other companies that had chosen to specialize in cars. When Aurora entered the field, it did so by way of a modest line of four trucks: Milk Truck, Gasoline Trailer Truck, Forty Ton Trailer with Power Shovel, and Twenty Ton Van Trailer (681-684). HMS gave the trucks innovative vinyl tires and some neat working features, but it turned out that kids wouldn't buy them. The trucks were advertised as 1/48 scale, but actually were sized a bit smaller than that. Aurora tried again, releasing them as Army trucks (304, 305, 306), molded in olive plastic, but then quickly dropped them from the sales list.

The car kits Aurora began selling in 1959 were much more successful. All of the first auto models were made from molds purchased from other companies, but the tooling had been originally designed by the craftsmen at HMS. Before coming to work for Aurora, Ray Haines had, of course, been working for his own company's interests. Among other companies HMS made patterns for was Advance Molding Corporation, a firm in New York City that manufactured a variety of plastic products. Haines had talked Advance into trying out the hobby market by selling a set of five sports cars, which they retailed from 1954 through 1956—then they decided to sell out to Aurora.

Among Aurora's tiny truck models, the most interesting is the Trailer With Power Shovel (683 $125-150).

The ex-Advance cars were very simple models, produced in inexpensive beryllium copper molds and lacking features like clear windshield parts and chrome plated bumpers. They did have black vinyl tires, stamped with the "Champion" company name, that fit into imitation wire wheels and snapped onto the chassis. Despite their shortcomings, these kits sold very well. At 49 cents these small 1/32 scale models were a good buy. Not only did they stay in the catalog into the late 1960s (1971 in the case of the Jaguar XK-120), but they were the basic models used in many of the "customized" kits released in the early 60s.

Advance Molding Corporation first sold a set of sports cars later taken over by Aurora. The Ferrari America (513 $40-50).

When the ex-Advance cars appeared in the 1959 catalog, another set of six "Indianapolis 500 Winners" were listed, too. These cars had been purchased from Best Plastics of Brooklyn, and their history almost exactly parallels the Advance cars. In 1953 Haines had gone to Best, which manufactured plastic birthday party favors, with the idea of joining the plastic model kit fad. The unoccupied spot in the industry he suggested to them was Indy race cars. To gather information for the models, Haines traveled to Indianapolis and watched the Memorial Day race, which was won that year by Bill Vukovich in the Fuel Injection Special (526). Haines collected data on Vukovich's car and then went to Speedway manager Wilbur Shaw, who had won the race in 1940, to get photographs of earlier winning cars.

Best started selling the car kits in the spring of 1954 (presenting gold plated copies to Shaw and Vukovich), and continued with the line until selling them to Aurora in late 1957. Aurora cataloged them from 1959 through 1963, but the kits were too primitive by the standards of the 1960s, and they were in an odd scale (1/30). This line was eliminated in 1964.

Another set of models Haines had created for Best came to Aurora with the cars. These were three "Historic Rifles." Haines fancied guns, as well as cars and planes, and put a good deal of care into these models. They were molded in three colors: black for the metal parts, dark brown for the stocks, and light brown for the gun racks. The rifles had spring-loaded working bolt actions. Unfortunately, they just didn't sell.

When Best got out of the model kit business, Aurora purchased their tooling and added Indy Cars to its inventory of models. The Maserati (525 $40-50) appeared in Aurora's trademark metallic burgundy plastic.

Antique Autos

The next group of automobiles Aurora brought out were 1/16 scale "Old Timers." Aurora got the name from Hudson Miniatures of Scranton, Pennsylvania. This company had been started by Anthony Koveleski, who once drove his prize Stutz Bearcat to the HIAA convention in Chicago in the dead of winter to publicize Hudson's antique car kits.

Old Time cars were popular subjects for model makers in the 1940s and 50s, and the Stanley Steamer (573 $30-35) tapped into this market.

Koveleski's original set of fourteen Old Timers were wooden, with metal and plastic detail parts. However, in the late 1950s he got out of the hobby business and sold all of his assets to Aurora. HMS picked out six of the Old Timers to recreate as all plastic models. The first car in the series, was Koveleski's Stutz Bearcat (571).

Aurora put a lot of effort into the Old Timers, packaging them in boxes with artistic turn-of-the-century designs. The kits were expensive to manufacture and included rubber-vinyl tires, metal plated parts, and molding in two colors of plastic. When *American Modeler* magazine asked Abe Shikes what his favorite model was, he picked the Stanley Steamer (573). "Even the name has a nostalgic ring to it!" This line of antique cars did not have great appeal to youngsters, who wanted modern sports cars, and many of the kits ended up in the hands of adult hobbyists.

The Old Timers pioneered an innovation in Aurora's car models: "chrome plated" or "metalized" parts. At first Aurora sent the parts to be "chrome plated" to Vacuum Metalizing Corporation in Brooklyn, but this made Aurora dependent upon another company for a critical element of its auto model production. As it happened, VMC went out of business, and Aurora hired its top two engineers to come to West Hempstead and set up a metalizing department.

By the early 1960s some big name companies were backing out of the plastic model business. Strombecker (Strombeck-Becker), a venerable hobby company that had been making wooden kits for years, ventured into plastics in the mid-1950s with only a few kits. Early in 1961 the company was bought by Dowst Manufacturers of Chicago, the maker of Tootsietoy diecast cars, and Strombecker abandoned models. Aurora picked up only two of Strombecker's molds, the Cessna T-37 (138) and the Temco TT-1 (139). In a sign of changing interests, Strombecker turned to concentrate its energies on a hot new item: slot cars.

An important fact stood behind Aurora's regular purchase of other companies' assets: Aurora had lots of cash on hand to invest in expansion. Indeed, Aurora's profit margin produced envy throughout the industry. Back in 1953, the first full year of model production, Aurora's total sales had been $1.3 million. By 1959 that figure had reached $5 million.

In 1959 Aurora gained a listing on the American Stock Exchange and made its first public offering of stock. The money from the stock sale went into debt retirement and purchase of new equipment. Although it did not seem particularly significant at the time, I. W. Burnham, a representative from the company that handled Aurora's Wall Street affairs, joined the board of directors. For the first time "outsiders" had a say in Aurora's affairs, but Shikes, Giammarino, and Cuomo remained the most important shareholders.

Why did Aurora earn such large profits? In great measure the answer was simple: old fashioned hard work and frugality. Each of the partners had lived through the Great Depression and treated money very carefully. Shikes wrangled with plastic supply companies like Monsanto and Dow for the lowest possible prices. Giammarino had a technician's thirst for efficiency. Wages at Aurora were lower than at many other major companies. The bosses switched off the lights when they left a room. Willem Thomas of Aurora Holland remembered leaving a party early with Shikes for a "midnight raid" on the third shift at the plant. They found everything in order, except for an unlocked door, and Shikes let everybody in the place know about it.

Everyone in management worked a twelve hour day. Sometimes Joe Giammarino became involved with a problem and simply worked through the night and into the next day. The assembly line workers noticed, and followed the pace set by their leaders.

As Henry Kolodkin put it: "It was a company that, right from the start, treated everybody so nicely. Everybody was treated as a human being. If you deserved something, you got it. You didn't have to ask for it. They made you feel you were part of a family. When Aurora made money, at the end of the year you got better bonuses." Aurora was a winner, and everyone knew it.

Chapter 3
The World's Largest

The most important molds Aurora ever bought from another company were purchased in 1956—and thrown right in the trash. The molds were those for three knights in armor, and despite their demise, they launched Aurora into the world of figure kits where the company would gain enduring fame.

The prototypes of the knights had been made by HMS at the request of Aaron Sloin of Princeton, who intended to market them through a new company, Crown Plastics. Sloin showed them at the HIAA convention in January, 1956 and took orders from distributors for delivery the next fall. At the time of the convention Sloin had not made the tooling because he first wanted to see whether the industry was receptive before he invested money in the molds.

With plenty of orders for kits in hand, Sloin took HMS's patterns to a mold maker who tried to build up a metal mold around the parts by electroplating them. This didn't work, so Haines introduced Sloin to Abe Shikes, and Aurora agreed to take over production and sale of Crown's knights. Sloin would receive a royalty on kits sold, and his Crown logo appeared on the first edition of Aurora's knights.

After Sloin's attempts at mold making were discarded, the patterns were shipped to Ferriot Brothers in Akron, Ohio, who normally produced the tooling used to manufacture toys and small play set figures like those sold by Lido, Marx, and Ideal. Ferriot Brothers did not cut their molds in steel because steel molds do not work well with figure models. Hard, precise steel molds serve well for airplanes and modern day ships with smooth, clean lines. However, steel is not a good medium for molds with many compound curved surfaces. Ferriot made its molds in beryllium copper alloy, a metal that is quite hard, but not as hard as steel. Molds made from beryllium copper were cast, not cut, as with steel molds. Ferriot would take one of HMS's patterns and make a plaster impression from it; then, using the lost wax method, cast the tool parts. This resulted in a mold that lacked the sharp, true lines of steel, but had texture and shape. This was essential in figure kits, very useful for some aircraft parts with complex curves, and produced realistic wood grain patterns in sailing ship models.

Since beryllium copper molds were also cheaper to make, Aurora sometimes made small scale auto kits from beryllium, rather than investing in more expensive steel tooling. Beryllium molds were quicker to manufacture than steel, taking only about three months, and a duplicate mold could be done in another five weeks.

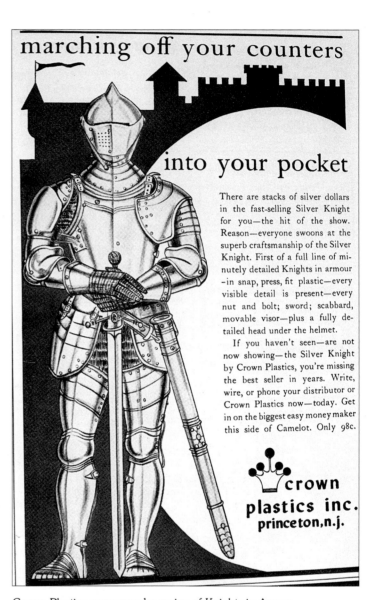

Crown Plastics announced a series of Knights in Armor models, but failed to bring any kits into production. Aurora took over the concept and made it a success.

Many molds contained both steel and beryllium parts. For example, the *Black Falcon*, Viking Ship, and Chinese Junk had beryllium copper mold parts for the hull and steel for the rest of the smaller parts. An airplane mold might be mostly steel, but some of the parts with complex curves would be beryllium. Telling which parts of a mold were steel and which were beryllium copper was as easy as distinguishing between a dime and a penny.

All in all, the beryllium copper molds were a fantastic deal for Aurora. An inexpensive mold could produce a model that sold tremendous numbers of kits in hobby shops. The prime example was Frankenstein (423), which at $6,150 was one of the cheapest molds Aurora ever bought, yet it produced millions of kits.

Figure Kits

Aurora's first figure kit, the Silver Knight (K-1) made it into the 1956 catalog, and the Blue Knight (K-2) and Black Knight (K-3) appeared in the 1957 catalog. Aurora added the Red Knight (K-4) and Gold Knight on horseback (K-5) in 1958 and 1959 respectively. The Green Knight of Landschut made a shadowy appearance in the 1957 catalog as a "knight to come" but never evolved beyond the pattern stage.

Aurora's knights sparked a boom in figure kits that swept the industry in the late 1950s. Having done planes, cars, and ships, companies jumped on the figure bandwagon. Bachmann sold animals, birds, and dogs; ITC did dogs; Palmer had big game trophy heads and soldiers; Precision had trophy heads; Revell produced Dr. Seuss characters; Pyro did dinosaurs—and there were more.

The bases of the Three Musketeers (K-8, K-9, K-10 $70-90) fit together into a semicircle. Aramis returned solo in the 1960s as D'Artagnan (410 $50-70).

kept a built-up on his desk. The horse also served as the mount for the Apache Warrior (401) and a cavalryman who was announced as "Jeb Stuart," but came out as the Confederate Raider (402).

All of these figure kits—with the exception of the Red Knight, the never-produced Green Knight, and the two gladiators—were created by one man, William J. Lemon. To the extent that Aurora's success depended upon its figure kits, Bill Lemon was its one indispensable man. Not only did he carve the patterns for the figure kits, he also sculpted the patterns for some of the pilot and ground crew figures in the 1/48 scale airplanes, and the plate designs for Coppercraft. Those who worked with him described him as a "totally incredible" sculptor. He could do human and animal fig-

The colorful knights in armor launched the figure kit craze and remained steady sellers for years. The Green Knight pattern was made, but Aurora decided not to bring it into mass production.

ures with natural features, proportions, and poses. He also worked very quickly, and he knew how to break down his sculptures into the model parts so that parting lines between pieces would not be obvious in the built model. The stiff, formal postures of Aurora's 1950s figure kits were not a result of any deficiency in Lemon's skills or limitations of the molding process—at the time no one thought of designing more dramatic action poses for the figures.

Aurora extended its figure kit line with the US Air Force Pilot, Sailor, Infantryman, and Marine (409, 410, 411, 412), the Three Musketeers (K8, K9, K10), the Viking (K6), the US Marshal (408), the Crusader (K7), and two Gladiators (405, 406). The horse from the Gold Knight was packaged separately as Black Fury (400), Aurora's first animal kit and, in the words of Kolodkin, "a tremendous seller." It was also a favorite of Joe Giammarino, who

Lemon started on the path to becoming a professional sculptor shortly after returning home from World War II. In 1949, while working as a cabinetmaker in Philadelphia, he answered a newspaper ad for a "model maker," having no idea just what that might be. He drove to the Philadelphia suburb of Bryn Mawr and climbed the steps to the workshop of Philip Derham & Associates above a movie theater. He found that Derham created the prototypes for all sorts of toys from cap guns to boats to airplanes to toy soldiers. One of the employees, Raymond Meyers, became Lemon's tutor. "I sort of took him under my wing," Meyers later recalled. "He'd ask me for advice, and I'd try to steer him in the right direction. He came along real fast—picked it up like a snap. He was a real natural." Fortunately, the assignments were not too complicated or demanding, and Lemon was able to learn his trade as he worked.

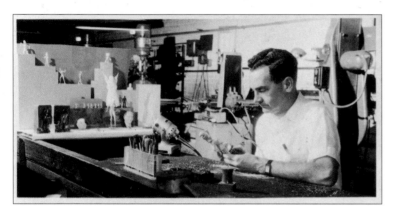

Bill Lemon sculpted alone in his home workshop.

Lemon had received no formal training in sculpting. "It was just something I picked up. I never graduated from high school. I never went to art school. . . . For some reason it came easy to me." He did his sculpting in acetate plastic. Lemon would begin with an array of small plastic blocks and a set of three-view drawings of the figure. He would look at the profile view of a part of the figure, and, as he explained, "I'd figure out—just guess— where the parting line would come." Then he would cut one of the blocks of plastic into two pieces with a band saw. That cut would become the parting line between two model pieces—for example, the front and back halves of a leg. Lemon would sculpt one side first, then, with that completed, he would sculpt the second side. If he had been careful not to let the acetate warp during the sculpting process, both sides fit back together perfectly since they had been cut from the same block of plastic.

The actual sculpting would be done with a rotary power tool, much like a dentist's drill. Lemon would grind the plastic into the approximate shape he wanted, producing clouds of plastic dust. Then he would begin shaving and scraping the plastic into more precise shapes with a variety of tools, some of which he had fashioned from bent and sharpened hacksaw blades. Finally, he would sand and polish the figure parts to a smooth finish, weld the pieces lightly together, and paint the completed figure. Ray Haines would take it to show the boys at West

Hempstead, and, once approved, the patterns would be photographed by the art department and sent to Ferriot Brothers to be made into molds.

The pattern for The Creature (426) was solid acetate plastic. The hollow cores would be created at the mold shop.

For years, sales manager John Cuomo had been requesting products with appeal for females. Lemon had sculpted bird and animal models that Bachman tried to sell to girls, and now Cuomo asked him to do a line of figure kits for Aurora that would have appeal in the "crafts" world of women. The result turned out to be Guys and Gals of All Nations—couples dressed in distinctive national costumes, standing on bases with maps of their country.

The first kits, the Dutch Boy and Girl (413, 414), appeared in the summer of 1957, and production went ahead on four other sets. Shikes, his wife Maria, and Cuomo held high hopes for the series—but Giammarino harbored doubts. As it turned out, the Guys and Gals did not take off as anticipated. In fact, a Swiss couple and a bride and groom set were killed after the patterns had been made.

Aurora built models for display in stores. The Chinese Mandarin, Dutch Boy, and Scotch Lassie (415, 413, 420 $20-30).

Each of the Guys and Gals (except for the Mexicans) came both as an individual kit and boxed as a gift set. *Photo by Pat Mazza.*

Although the Guys and Gals did not make money, they brought a man to Aurora who would have tremendous impact on the company. Donald "Bill" Silverstein had been raised in New York and grew up with a young man's dream of playing professional baseball. He had signed with the Giants when World War II intervened, and he joined the Army Air Corps, flying in B-17s over Europe. On his 18th mission his formation of bombers was jumped by ME-109s. Silverstein bailed out and survived a year in a German POW camp only to nearly be killed by American P-51 fighters strafing his column of marching prisoners during the last days of the war. Back in the United States, he got his chance to play minor league ball, but ended up in the advertising game with Harold J. Siesel Company.

The Siesel agency put him to work on hobby accounts. He went to Renwal and helped them advertise their Visible Man figure kit. About the same time, he went to see John Cuomo, who wanted him to develop the kind of promotional program that Revell had long used. Aurora had a reputation for being very tight with its advertising dollar and was still not willing to invest much more money in advertising, but Cuomo expected Silverstein to make the most of his resources. He moved into the same small corner office occupied by the company's three founders. "When I came, the three of them worked out of one room. They opened the mail themselves. They had no secre-

tary and no receptionist to answer the phone." Silverstein pushed his desk into the unoccupied fourth corner of the office.

His first project was a sales campaign for the Guys and Gals, a name Silverstein considered "dreadful." The first component of the promotion was a full page, color ad in the March, 1958 issue of *Parents Magazine*, which described the figures as having "the same elaborate detail as a Dresden figurine." Displays, along with factory built-ups of the kits, were sent to hobby shops. As a side benefit to this campaign, Aurora gained permission to use the *Parents Magazine* seal on any of its products, and kits from the late 1950s are likely to have the Parents seal on their boxes and instruction sheets.

By 1960, however, the figure kit fad was faltering. William Lester, president of Pyro, said that his company had run trials with average model builders, who found working with figure kits frustrating and the results disappointing. The big problem turned out to be painting, since few kids had the skills to paint the eyes and other details.

Companies started to drop figure kits. Aurora's 1959 catalog contained thirty figure kits; its 1960 catalog had only eleven. The major casualties of this massacre were the Guys and Gals. The knights survived and would continue as steady sellers because they looked neat with little or no painting.

After the Guys and Gals proved to be mediocre products in the marketplace, Aurora cancelled further development of the Swiss boy and girl, as well as the bride and groom.

An area that looked very promising for new kits in the late 1950s was rocketry. Following Russia's launching of Sputnik in the fall of 1957, Americans were caught up following the "space race" with the Soviets. Scenes of rocket tests appeared on TV almost every night. Aurora jumped into rocketry with a battery of eight "news-making missiles" in 1959—but the series crashed and burned. Kids would not buy models of pilotless, impersonal machines. The missiles were carried in the 1960 catalog only to help move the stacked-up inventory.

One way to find out what products customers will buy is to ask them. So in 1959, Bill Silverstein launched Aurora's "Dream Kit" contest. Entrants were asked to send in their concept of a new kit, in twenty-five words or less, and the winner would receive a Czechoslovakian-built Skoda 440 America sports car. There were other prizes, from motor boats to Capitol phonograph records. The catch was that you had to send in a box end with each entry. What Silverstein primarily wanted from the contest was publicity and a direct boost to sales, not the ideas. The entries did pour in—cartons and cartons of them, stacked up in the hallways and most never read.

The model suggestions did reveal a couple of interesting things. Entrants confirmed that they didn't care about rockets, but, surprisingly, figure kits generated lots of enthusiasm. The most commonly requested figure models were religious personalities—although Brigitte Bardot got many votes. Also mentioned were every car, ship, airplane, and army vehicle ever made. Aurora's staff had already thought of these. Some of the weirder suggestions were for things like one-eyed blob monsters. Now there was some idea—monsters.

The official winner of the contest was sixteen-year-old Craig Brun of Rochester, New York, who flew to New York City with his dad to pick up the Skoda (which turned out to be a nifty car). His winning idea, a set of scale models of Washington's public monuments and buildings, was not announced. Privately Silverstein acknowledged: "There wasn't any entry that we ever made a model of."

As a publicity stunt, the Dream Kit contest emerged as a smashing success. Aurora had gotten a lot of mileage from a small investment of money and effort.

Silverstein's next promotion turned out to be equally successful. In the September, 1959 issue of the toy industry's trade journal *Playthings*, Aurora ran a two-page spread proclaiming that Aurora already had the "world's largest collection of plastic scale models" (a claim based simply upon Silverstein's count of the kits listed in the Aurora and Revell catalogs—Aurora had 152, Revell 118). But more than that, the ad announced "The crash program that establishes Aurora as America's No. 1 maker of plastic hobby kits!" (Revell responded to Aurora's rumblings about being number one by releasing its own survey of hobby shops claiming that Revell outsold the next three model companies combined. However, if dime store sales had been included in the totals, Aurora would have made a much better showing.)

When the brief boom in missile kits sputtered out, Aurora dropped plans to bring out models of the Mace and Honest John.

The Apache Warrior (401 $300-350) is one of the rarest 1950s kits. *Photo by Pat Mazza.*

The program that supposedly would make Aurora number one was a premium offer advertised in a ten page ad in the December issue of *Boys' Life*. By sending in two box ends and some money, boys could buy Eveready flashlights, Rawlings sports equipment, watches, knives, and ball point pens. In May, Aurora ran another ad for Fathers Day, offering cuff links for box ends and $2. In December another eight page ad appeared in *Boys' Life* with the same items available.

However, in the spring of 1961 Abe Shikes killed the promotion, declaring, "We are in the plastic business." It had been quite rewarding and cost Aurora nothing but the effort involved because Aurora had sold the premium items for exactly what it cost to buy them from manufacturers.

Aurora and other model companies also promoted kits by tie-ins with movies. In 1958 Aurora linked two of its kits, the Viking Ship (320) and the Viking (K-6), to Kirk Douglas's movie *The Vikings*. Douglas posed for a publicity shot with the Viking Ship, which was touted as "an exact reproduction of the ship used in the movie." (It wasn't.) Paper shelf streamers were mailed to hobby shops with pictures of Aurora's Viking Ship and actors from the movie. The Viking figure kit was modeled on Douglas himself—although the model's clothes do not match those in the movie. "I tried to get a likeness of him," said sculptor Bill Lemon. He got the chin dimple!

When the remake of the old Cecil B. DeMille movie *The Buccaneer* came out in 1958, Aurora had artist Jo Kotula paint new box art for the *Black Falcon* pirate ship and issued it under the name *The Buccaneer* (429). In 1960 Aurora released the aircraft carrier *Enterprise* and the Japanese battle-ship *Yamato* kits to coincide with the premier of *The Gallant Hours*, a movie about the Battle of the Philippine Sea in which both ships took part.

Sometime in 1957 Jo Kotula got a phone call in his Morristown, New Jersey studio. It was Aurora's printer George Burt: "How would you like to do a painting a week for the rest of your life for Aurora?" At the time Kotula was the most noted aviation artist in the country, having done covers for *Model Airplane News* since 1931. Burt admired his work and had taken one of Kotula's magazine covers, wrapped it around a kit box, and showed it to the men at West Hempstead. They liked the impressionistic style of Kotula's work. "The editor of *Model Airplane News* had asked me for that brilliant sunset quality… I tried even for some jarring color contrasts on the plane because that was our game—to catch the eye on the news stands." Kotula visited the Aurora plant, and got another suggestion from Joe Giammarino: "When I see the box, I want to get the feeling that I want to duck!"

Jo Kotula's art explodes with color and action. Typically, as with the F-86D (77), the illustration on the box top was better than the model inside the box. *Photo by Rick DeMeis.*

Kotula came on board at Aurora as their primary artist, although all his work was paid for on a freelance basis. He would travel to Long Island for some of the lunchtime meetings and discuss box art with Giammarino, Burt, and the sales staff. For the most part, Aurora gave him a free hand to do as he wished, working only from standard box dimensions and photographs of the model prototypes. He painted in opaque watercolors, gouache, a medium favored by commercial artists because it is easy to work with and contains a lot of white in its base that reflects well when photographed for reproduction. "It gave it a more meaty, more realistic, more gutsy quality," declared Kotula. He did not attempt to capture a photographic reproduction of the airplane, but its essence. In this respect, his art paralleled Aurora's modeling philosophy.

Aurora used Kirk Douglas's movie *The Vikings* to promote sale of its Viking figure model and Viking Ship (K-6 $110-130, 320 $25-65).

A common Kotula composition would show two images of the same aircraft. The Curtiss P-40 (44) made a great subject for a painting.

In all, Kotula did more than eighty box wraps for Aurora. His first was the M-2 Mail Plane (111) in 1957, and his last were the four World War I kits released in 1964. He also did some of the chalk drawings which appear in the annual catalogs. For a routine painting, he would be paid $350, but for his most elaborate wrap, the World War II carrier *Enterprise* (714), he was paid $1,200. On some of his aircraft paintings, Kotula was assisted by Keith Ferris, then just a teenager, who drew the pencil outline of the plane for Kotula to work from. Ferris would go on to become a noted aviation artist in his own right.

The USS *Enterprise* (714) was Jo Kotula's most elaborate painting.

Kotula did not want to be labeled as just an airplane painter and often included human figures in his foregrounds. Several of Aurora's figure kit boxes were done by Kotula, including the Gladiators (405, 406), the Apache Warrior (401), and the Confederate Raider (402), the one Aurora painting which Kotula kept for himself after Aurora copied it. Also, he painted some military vehicles and ships. Giammarino did not like Kotula's ship pictures. He criticized Kotula's Bluenose Schooner (431) as too "antiseptic" (yet Giammarino took that painting for himself and hung it in his summer home in Connecticut).

When Aurora began to slow down production of new model airplane kits in the 1960s, Kotula looked elsewhere for commissions. The Air Lines model company hired him to do box wraps, and the Air Force asked him to illustrate some of their publications. Kotula would continue right on painting, including a 1990 cover for *Model Airplane News* commemorating his nearly fifty years of work.

Kotula was an artist with varied skills who also painted portraits and landscapes, and he felt that including human figures in model kit art added an element of personal interest. The Flying Boxcar (393).

New Hobby Fields

As Aurora grew larger and more prosperous, and as growth in the plastic model kit market seemed to be stalling, Aurora experimented with a variety of new products that might broaden its opportunities for profits. In 1959 it brought out a new product aimed at the toy market: the Mark III Interceptor Missile. Aurora intended to cash in on the popularity Amsco Industries of Pennsylvania was enjoying with its Alpha I rocket. Both missiles were revivals of an old idea—mix citric acid, bicarbonate of soda and water—and you get a zesty fizz powerful enough to send a toy rocket flying. Shikes got Giammarino to design a launch pad that clamped the rocket in place while pressure built up inside the rocket's mixing chamber. Then a yank on the launching string would pull the plug in the exhaust hole and send the missile skyward.

The Interceptor Missile resulted in two law suits. First, the maker of the Alpha I sued Aurora for infringement on their product. However, Amsco didn't have a copyright, and the judge dismissed their suit, saying a the rocket concept was public property. The second suit was more serious. A young boy was shot in the eye by an Interceptor Missile fired by his father. Aurora's insurance company handled the settlement, but Giammarino killed the missile, saying, "I don't want that damn thing to hurt children."

Another toy sold well for a few months, but turned out to be another flash-in-the-pan. The "Bloop Gun" consisted of a plastic tube with a pump handle on one end that could be pumped-up to shoot fat plastic shells. Shikes bought the idea from Chicago toy designer Marvin Glass (one of the nation's all-time great toy innovators), and HMS engineered the product. It was marketed by Aurora as the "Tank Buster," but Silverstein thought it would sell better as a lighthearted toy and dubbed it the Bloop Gun. He made Aurora's first TV advertisement and aired it in the New York City area during the spring of 1960. The TV ad caught school boys home on Easter break, and for a few months Aurora had two molding machines working to supply retailers in New York City, but orders soon evaporated, and so did the Bloop Gun.

Aurora had one more product that verged over into the toy realm: motorized model cars. Moving parts and other "action" features were a fad in plastic models at the time, and several companies put electric motors into their models. Aurora sold a pair of cars that were scaled up to a large 1/11 size so that they could accommodate two D batteries and an electric motor. The Pontiac Fireball (531) and Ford Hot Rod (532) lacked the detail and accuracy of models, although they did have lighted headlights, transparent red tail lights, and black plastic tires. They were intended to be assembled and then played with as toys. The problem was that the motors and electric features made the kits expensive, and once you built the cars they lacked "play value" because they could be set to run only in a straight line or a circle.

Aurora had been induced to produce these cars by Nat Polk, whose company imported electric motors from the Orient for all sorts of hobby applications. Polk even talked Shikes into bringing out a set of electric vehicles that made only the slightest pretense to being hobby items. The "Go Toys" were blister-packed on a card and required only minimal snap-together assembly. There was a hot rod, a delivery truck, an Indy car, and a diesel locomotive. However, these pieces, and the other toys, proved hard to market to distributors, who knew Aurora for its hobby kits, not its toys. Ray Haines recalled Giammarino telling Shikes, "We're a hobby company, not a toy company." At the time Giammarino won the argument, but the question of entering the toy market would rise again a decade later—with major consequences for Aurora.

The Mark III Interceptor Missile was an attempt to broaden Aurora's presence in hobby shops.

Within the hobby arena, Aurora enjoyed much more success. In 1957 Aurora purchased Air Champ Radio and brought its owner Frank Carver in as production manager and chief procurer of parts and materials. Air Champ had a line of crystal radios and telegraph sets that it had successfully marketed for years, chiefly to the Boy Scouts and Cub Scouts. To cash in on the space race craze after the launching of Sputnik, Aurora put one of the crystal sets into a plastic case shaped like an earth satellite.

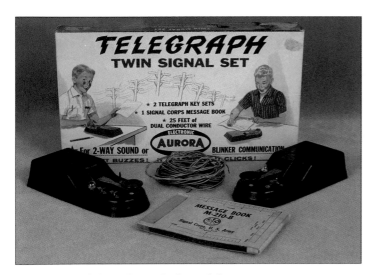

Aurora entered the radio and telegraph business when it purchased the Air Champ company. Aurora sold many telegraph sets to the Boy Scouts.

The Satellite Radio was the most interesting of Aurora's crystal sets.

Another hobby field Aurora plunged into was flying model airplanes, which had formed the core of the hobby industry since the 1930s and 1940s. Ray Haines and Mattie Sullivan had long been flying model enthusiasts. Haines designed a prototype flier made of sturdy nylon and plastic. He demonstrated it to Shikes and Giammarino in the parking lot at West Hempstead, crashing it into a mud bank to show that a well-built plane could survive even a severe crack-up and fly again. Haines powered his prototype with a simple motor designed by K & B Allyn, an old-time leader in gas engine production.

Aurora's flying models came in boxes that looked like airplane hangers.

The boys liked what they saw, and approved an L-19 Bird Dog and P-47 Thunderbolt. When sales of these planes looked promising, Fokker D-7 and SE-5 biplanes followed, accompanied by a trio of air racers. Then Aurora offered a big, twin-engine B-25 bomber. Although the mechanically-minded Haines and Giammarino loved the flying planes, others at Aurora had their reservations. Frank Carver noted the difficulty of procuring and assembling all the various parts that went into the planes. Customers who bought the planes often brought them back to the store for a refund when they couldn't make them fly. Abe Shikes cut a finger trying to get one started. Thus the flying model airplanes quickly fell out of favor in West Hempstead.

Production tooling was made for the Messerschmitt Me-109, but it never made it to stores as a flying model, and later in the 1960s it failed to join the "screwdriver" non-motorized model kits.

Ray Haines proudly surveys the production line for flying models.

However, the planes did have a lasting impact on Aurora: they brought K & B into the corporate fold. In order to insure a steady supply of engines, and to extend its range in hobbies, Aurora purchased K & B in February, 1960. John Brodbeck, president of K & B, came with the company and continued day-to-day operations.

John Brodbeck and Ludwig Kading had founded K & B back in 1944 in Compton, California. A slender man with dark, curly hair and a pencil-thin mustache, he had picked up his education in mechanics on the job. In 1955 Brodbeck merged his company with Allyn Sales, a company owned by Allyn Gasdia. Allyn also manufactured flying aircraft and motors, but it is best known to collectors today for its very rare plastic models of Douglas, Boeing, and Convair aircraft. These kits continued to be made until about 1958, when Gasdia suffered a heart attack and sold his portion of the company to Brodbeck.

One of the first things Aurora did with K & B was to move it to a large new plant in Downey, a Los Angeles suburb. Although K & B's main products were still connected with flying models, Aurora shifted some of its product lines to Downey to make room for new items back at West Hempstead. The two main products sent west were the flying airplanes and the crystal radio sets. Both were marketed under the K & B label through the mid-1960s.

About the time that Aurora got involved in flying airplanes, it also entered the field of model railroading by issuing HO scale buildings for layout landscapes. HMS had experience in designing model railroad buildings, having done some of Bachmann's Plasticville kits. Aurora released eight models, including Joe's Diner (657), named for Joe Giammarino. These were very good models, each molded in two colors, with features such as doors that opened.

HMS made patterns for ten other buildings, but these were never issued because the HO train craze leveled off in the 1960s. Among the kits not issued were the "Aurtown" factory building, a firehouse, a dairy bar, and a building with a hardware store and hobby shop.

In the summer of 1962, Aurora expanded its position in model railroading by purchasing a companion company to K & B in the Los Angeles area: Tru-Scale Models, a manufacturer of model railroading equipment and supplies. When Aurora purchased Tru-Scale, it sent the plastic HO buildings to California to be marketed. Also going to Tru-Scale were the Coppercraft kits Aurora had introduced a few years earlier. Tru-Scale continued to sell the HO plastic houses and Coppercraft as late at December 1965, but Aurora's partnership with Tru-Scale did not survive past the mid-60s.

Up to this point, none of Aurora's ventures outside the realm of plastic model kits had paid off in a big way; however, that was about to change. Their next gamble would transform the company.

The Gas Station (658 $150-170) is the rarest of Aurora's HO train and slot car structure kits.

Chapter 4
Movie Monsters

The March 1960 issue of the toy industry trade magazine *Playthings* carried a teaser advertisement by Aurora announcing, "An unbelievable new product destined to take its place among the best-selling items of all time!" Aurora introduced "Model Motoring," an electric-powered HO slot car system that became the smash sensation of the 1960s, leading Aurora to unprecedented growth that would soon make it the leading hobby company in the world.

Aurora's new plastic kit offerings, by contrast, went through some lean months in 1960 and 1961. A few new ships, planes, and cars were issued, but nothing in the way of a major new series. Plastic kit sales were not increasing at nearly the pace of slot cars.

The *Complete Book of Plastic Model Kits* came with the Aurora logo for sale in hobby shops and with the Dell trademark for sale in book stores. It covered only Aurora models!

Advertising and idea man, Bill Silverstein wanted to make the most of what Aurora had to offer, and he suggested repackaging the kit line to modernize its appearance. Central parts of his proposal were ditching the oval logo and returning to a rectangular trademark that he felt would better fit on boxes and in advertisements. However, he could not convince Shikes, Cuomo, and Giammarino to abandon the oval: "They loved it." He did get them to agree to drop the "Famous Fighters" words from the border and to "clean it up" by eliminating the sun rays in the interior of the oval. The old-fashioned practice of wrapping the box cover art around the sides of the box also stopped, and box sides became bold solid colors with the name of the kit repeated so that it would be visible no matter how the box was stacked on the shelf.

One day as Silverstein walked down the street, he spotted a crowd of youngsters waiting to get into a movie house. He went over and discovered that the theater was showing a Frankenstein-Dracula double feature. The movies were the old 1930s Universal Pictures that had been re-released in the late 1950s and were common fare on weekend television. Here was an idea for something new in plastic kits. "Remembering how, as a kid, the monsters thrilled me, I instinctively felt today's kids would go for them."

Silverstein went to Aurora's art director, Si Friedman, and had him make up a dummy box for Frankenstein's monster. Then he took his box and his idea to the next deli lunch meeting of the Aurora staff. Why not, he announced, make a series of all the monsters from the old Universal movies. The kids loved the movies, and it would give Aurora something different from all the other plastic kit companies.

The rest of the men in the room either laughed or moaned "No!" in chorus. Everyone knew what a disaster figure kits had been. Besides, explained John Cuomo, Aurora could not sell such gruesome stuff to children. Shikes felt they were sure losers. Aurora's retailing consultant, Dick Schwarzchild declared: "Forget it. That's crazy. No hobby dealer in the United States will carry a kit of Frankenstein."

The Big Horn Sheep and the Cougar (454 $120-140, 453 $110-130) are the most complex of the Wildlife Series kits, and they are the most rare today because they were discontinued in the mid-1960s.

Each week Silverstein would return with his Frankenstein box, until it became an in-house joke. Finally Silverstein issued an ultimatum: either make the monster kits, or he would resign and form a new company to do them. Shikes and Giammarino reluctantly agreed: Aurora would make one monster, Frankenstein, and see the reaction within the industry. Abe Shikes later admitted that he went along with the crazy monster idea just to give Silverstein the chance to make a mistake.

Silverstein then went to see the agent who owned the rights to Universal's movies. He quietly listened as the agent told about the movie and TV personalities he held the rights to. Silverstein then asked if he also had the monsters. "Sure," he replied, "Why would you want them?" Silverstein said he would like the rights to all the monsters, and would pay a three percent royalty on sales (this later increased to five percent).

Bill Lemon got the assignment to do the pattern for Frankenstein. Aurora gave him a toy Frankenstein—which could only have been Marx's wind-up walker—and asked him to turn it into a model. Lemon did his own design and working sketches for the figure and base. "I'm no artist," he explained, "but I could make a workable drawing." Once Frankenstein was done, Lemon delivered it to Aurora and waited to see what would happen. He had to wait for a long time.

In January 1962 the Aurora boys flew out to Chicago for the HIAA annual convention—except for Silverstein, who took a train because he had had some bad experiences in airplanes!

Aurora displayed Lemon's Frankenstein model pattern in its display booth, and the reaction of the industry was: "Aurora has gone mad." No one showed any interest in placing orders. John Cuomo used his customary line: "Try one order. If its a success, great. If not, you haven't lost much." There were no takers. On the final day of the convention Silverstein showed Frankenstein to one last distributor, and he declined to place an order. Silverstein sat down next to Cuomo, and John said, "Now will you forget it?"

The Bull Elk would have made a magnificent model kit, but it did not make it further than the pattern stage.

One of the references Bill Lemon used in sculpting Frankenstein's monster (423 $200-250) was Marx's wind-up walker toy.

The Wildlife series appealed to enough model builders to keep them in the catalog year after year.

On the last day of the show, industry representatives were allowed to bring their families into the show parlor, and on that day Chicago distributor Al Davis's sons Glenn and Fred wandered into Aurora's exhibit. Silverstein and Cuomo expected the boys to go for the Model Motoring layout, but instead they grabbed Frankenstein. The elder Davis was impressed with Aurora's willingness to try out the monster idea. He later declared: "The men at Aurora didn't pussyfoot around. They had the guts to do what others wouldn't do."

A couple of California distributors witnessed the episode with the Davis boys, and placed two orders for Frankenstein's monster. Aurora shipped out the kits later that year, and shortly thereafter the phone began to ring: "Can we have some more monsters?"

When it became evident that Frankenstein was going to be a fantastic success, Aurora had HMS's design team rush the rest of the line into production. The next two kits, Dracula (424) and the Wolf Man (425) were in stores just before Christmas, 1962. In 1963 the Creature from the Black Lagoon (426), the Mummy (427), and the Phantom of the Opera (428)

were released. The following year the Hunchback of Notre Dame (461), King Kong (468), and Godzilla (469) lumbered into hobby shops. Dr Jekyll as Mr. Hyde (460) did not reach the stores until 1965.

All of these monsters (except King Kong, Godzilla, and Dr. Jekyll) were created by Bill Lemon. By 1962 Lemon had left HMS and set up his own studio in Ambler, Pennsylvania. However, he still worked freelance for HMS. For the monsters, Aurora gave him stacks of 8" x 10" Universal Pictures still photographs to work from. Sometimes he also received three-view drawings of the design for the model. Lemon thoroughly enjoyed creating the monsters—except for the Phantom of the Opera. "That was one I always didn't like. That was one I thought was horrible. Pretty gruesome."

It took him only 75 to 120 hours to make a figure pattern, for which he would charge Aurora only $2,100 or a little more. Considering the low cost of Lemon's patterns and the relatively low cost of Ferriot's copper molds, the renewed interest in figure kits meant big profits for Aurora.

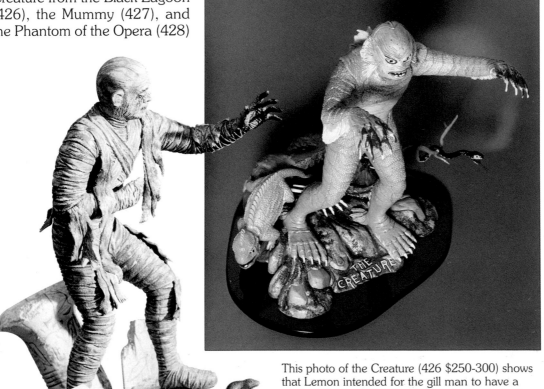

This photo of the Creature (426 $250-300) shows that Lemon intended for the gill man to have a crest on the top of his head. The rise in the back part of the head on the model was not caused by a defect in the mold, although the mold exaggerates the crest.

Lemon added some blood to his prototype of the Mummy (427 $175-200) to give it a little color.

Despite the success of Aurora's monsters, Lemon found himself deluged with orders from other companies for his services. Companies like Knickerbocker, Remco, and Marx monopolized his time with orders for toy figures. (However, he did not sculpt the well-known Marx movie monsters.) Thus he was not available to create any more monsters for Aurora after the Hunchback. His later 1960s production for Aurora included only Zorro (801), Superman (462), and four Wildlife kits: the White Tail Deer (403), Black Bear (407), Cougar (453), and Big Horn Sheep (454).

When Lemon departed, one of the men who stepped in to fill the gap was Raymond Meyers, Lemon's old mentor at Phil Derham's. Lemon felt that Aurora was in good hands: "Ray was terrific. He was the best." Although they were in competition for the same commissions, Lemon and Meyers maintained a close friendship. There was always plenty of work to go around.

Meyers had grown up wanting to be an artist. His mother realized his potential and encouraged it. "I guess I just had a talent in me. When I was a kid I was always drawing. Mom would buy me cakes of Ivory Soap, and I

would carve it. The more I did, the more I liked it." After World War II he found work sculpting prototypes for all sorts of household items—his first assignment was a coffee pot handle. Once he started work for Derham, he progressed to doing toys, Plasticville houses and farm animals for Bachmann, and animal model kits for Precision Plastics. The first assignment Aurora brought to Meyers was Godzilla.

The sculptors who created Aurora's figure kits also made hundreds of toy characters for other companies. Bill Lemon carved astronauts and other little people for Louis Marx. Ray Meyers did many push-puppets for Kohner Brothers, and Larry Ehling made early Disneykin-like figures back in the 1950s.

The Phantom of the Opera (428 $200-250) is the only monster to have a victim included as part of the model.

Aurora regularly sent out window streamers to hobby shops announcing its new kit releases.

Not surprisingly, Meyers used the same sculpting technique as Lemon. He also worked in acetate plastic. "It was beautiful to work with. It was like wood, but without grain marks. If you made a mistake, you could just cut it off, soak two pieces in acetone, bond them together, and start again." The most difficult aspect of model kit making was insuring a precise part fit. "Making the figurine was easy. Making the parts fit together was the most time consuming thing."

Time was a precious commodity for Meyers. Shortly after making the Godzilla pattern for Aurora, he left Derham and went to work as a freelance artist, working out of the basement of his home in Philadelphia. Kohner Brothers, manufacturers of toys for small children, hired Meyers to do a lot of work for them. "They wanted me to work eighty hours a week. I didn't have time to do anything else." Thus pattern making for Aurora had to be squeezed into available time.

Meyers knew that sometimes when Aurora had two models that were supposed to go together as a pair, each one would be allotted to a different sculptor. That way both could be completed more quickly and at about the same time. Thus it did not surprise him that someone else was asked to do the mate to Godzilla—King Kong. This task went to another old associate at Derham, Adam "Larry" Ehling.

Ehling was considered the "old man" at the shop. Although Philadelphia-born, he loved the wild west and lived in a rustic log cabin that he and his wife built in the countryside. Ehling had grown up in the old handicrafts tradition of wood carving and liked to make models of covered wagons, as well as figures of cowboys and Indians. Unlike Lemon and Meyers, who went off to work on their own, Ehling remained at Derham through most of career. When neither Lemon nor Meyers were available to make a pattern for Aurora, the task often fell to Ehling.

The King Kong (468 $250-350) pattern was probably sculpted by Larry Ehling.

Godzilla (469 $250-350) was the first figure kit Ray Meyers did for Aurora—and among his favorites.

The most unusual monster to debut in 1964 was Gigantic Frankenstein (470), a nineteen inch tall caricature of the movie monster. Outside toy designer Reuben Klamer had originally conceived it as a battery-powered walking toy, but it became a $4.98 model kit. Unlike the other monsters, which maintained a dignified scariness, Big Frankie played the role of parody. He lasted only two years in the catalog.

During the monster craze of the mid-1960s, Aurora brought out Coppersmith kits of Frankenstein, Dracula, the Wolf Man, and the Creature.

41

The demand for monster kits became so great that Aurora went to unusual lengths to produce them. Duplicate molds were made of Frankenstein and Dracula. Frank Carver remembered, "They never came off the machines. We used to bang 'em, bang 'em, bang 'em every day of the week." Some of the molds, after being thoroughly tried out at Aurora's plant, were sent out to contract mold shops to produce monster parts. (In all, Aurora had perhaps thirty "vendors" doing contract molding on various items, not just the monsters.) By the start of 1964 Aurora had sold seven and a half million monster kits.

In the fall of 1963, Aurora began an advertising program that boosted monster sales even more and changed the course of Aurora's advertising. A representative from National Comics Group approached Silverstein with an offer of low cost advertisements on the back cover of DC comic books. DC had recently lost one of its standard advertisers, Daisy air rifles, and was looking for a new client. The price was so low for this twelve month package that Silverstein jumped at the offer.

The state of New York told Aurora's advertising head Bill Silverstein that its "Everything you need to make your models come alive" slogan was false advertising. Silverstein replied: "You've got to be kidding!" But they weren't. Package art for the Monster Colors paint set was done by Mort Künstler.

Beginning with the November, 1963 issues, DC comics ran full color back cover ads for Aurora products. DC sold about seven million comic books a month at the time, and comic books stayed around in kids' rooms for months, giving the ads repeated exposure. The response to the DC back cover ads turned out to be so good that the next year Aurora had to start sharing the cover with other companies, and some Aurora ads were pushed to the inside pages. Nevertheless, Aurora and DC had established a relationship that would last for years and lead to a new line of Aurora comic book hero kits.

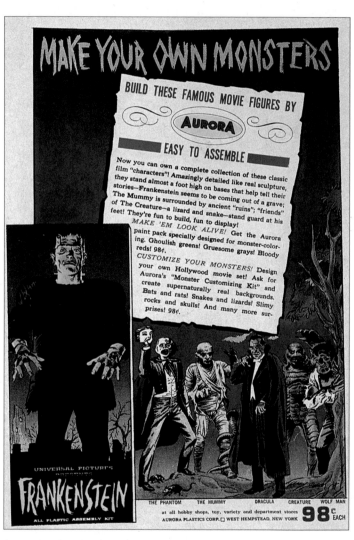

Aurora's advertising on the back covers of DC comics began with this ad.

In 1964 Aurora sponsored the Monster Customizing Contest in cooperation with Universal Pictures and James Warren's *Famous Monsters of Filmland* magazine. Beginning in January, Aurora sent contest packets to hobby shops around the country containing entry blanks, window streamers, "Master Monster Maker" certificates for entrants, and plaques for store contest winners. Entrants were allowed to use Aurora models in any combination and with whatever customizing materials they chose. Two Monster

Customizing kits (463, 464) were released to provide rats, bats, spiders, skulls, and other assorted scary stuff for improving the monster kits. State winners were selected from photographs of store champions' creations, and national finalists sent their complete projects in to Aurora.

Silverstein described the entries as "magnificent, beautiful—well, *macabre*." Some of the youthful monster makers dressed up their creations in real fur and cloth clothing. All the finalists were required to place their models in elaborate "horror movie sets." National winner Greg Gellman of Oklahoma City built a two-story House of Horrors with a mad scientist, a suit of armor upstairs, and Frankenstein's monster on a lab table downstairs.

Aurora's monster kits were the catalyst for a monster craze that swept the nation in 1964. That year Frankenstein, Dracula, the Wolf Man, and the Mummy were almost as much of a mania as John, Paul, George, and Ringo. All sorts of monster toys, games, and puzzles hit the market. *Life*, *Look,* and *Newsweek* magazines carried stories on monster mania. In the fall CBS began broadcasting *The Munsters*, and ABC premiered *The Addams Family*.

Other model companies edged onto the playing field with kits that were offbeat, but not quite monsters. Revell had Rat Fink characters driving race cars, Hawk made Weird-Ohs personalities, and Lindberg arrived on the scene with Loony Repulsives. Silverstein thought Aurora's kits were the best, boasting, "We won't make any fly by night monsters. We stick with the classics."

Entrants in Aurora's Monster Customizing Contest received Master Monster Maker certificates.

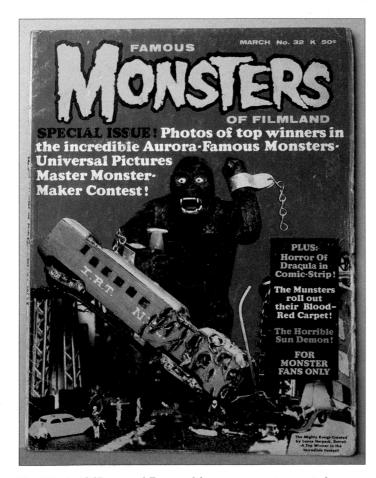

The March 1965 issue of *Famous Monsters* magazine carried a 12-page story on the Monster Customizing Contest.

The Addams Family Haunted House (805 $400-500) had a short life on store shelves in the mid-1960s. Polar Lights picked it as its first Aurora kit to revive in the 1990s.

The Trouble with Monsters

Aurora found itself deluged with protest letters from outraged parents fearful of what these monstrous models were doing to their children. Anticipating this reaction,

Aurora had gone to professional psychologists to find a defense: "In the opinion of reputable authorities, movie monsters actually perform a valuable service to the child. Certain fantasies are harmful only if improperly focused. When they center about an imaginary object, like a picture, a movie, or a model, they are released in the manner of steam, escaping through the safety valve of a radiator." Silverstein was pleased to recall one letter from a mom who said that her son had been troubled by nightmares until he started building little monsters that he could control.

In August of 1964 Steve Allen's nationally syndicated TV talk show took up the monster question. Allen, his wife Audrey Meadows, a psychologist, and a few Boy Scouts sat behind a coffee table with Dracula, the Creature, and Revell's Rat Fink on display. During the show the Boy Scouts said that they talked about monsters all the time, the psychologist called the models "play therapy," and audience applause favored the monsters over the anti-monster critics. But the monsters did not have it all their way. One guest picked up the Creature and said, "Why cultivate something as idiotic as this when there are more beautiful subjects to draw from." (Clearly, this panelist was unfamiliar with the public's indifferent response to the wholesome Guys and Gals of All Nations!) Allen agreed that promoting beauty was desirable, and at the end of the program picked up Dracula and read, "Dracula, made by Aurora Plastics Corporation."

This show caused a furor back at West Hempstead. There the program was interpreted as a criticism of Aurora's products. The reaction was particularly strong because the show had been the idea of a publicity committee funded by the Hobby Industry Association. John Cuomo fired off a letter to the HIAA resigning from its board.

Aurora received more TV attention when CBS Evening News visited Aurora consultant Rich Palmer in his hobby shop in Parsippany, New Jersey on Christmas Eve. Walter Cronkite and Palmer held a conversation in the "Monster Korner" of the store. Naturally, Palmer spoke up for the monsters and for Aurora.

1964's Monster Customizing Contest directly influenced the next two Aurora kits issued in 1965, the Bride of Frankenstein and The Witch (482-483). Both kits featured

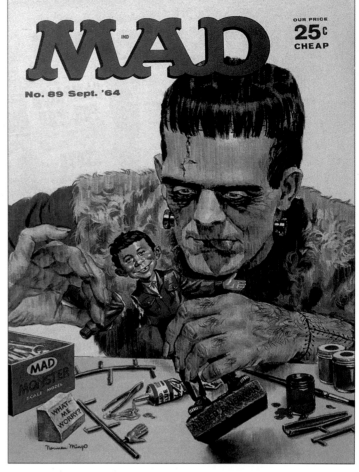

MAD magazine had fun with the monster-building hysteria. Aurora was one of the few companies that received permission to sell a MAD product: Alfred E. Neuman (802 $70-80).

diorama settings and shrank the figures to about 1/11 scale, instead of the 1/8 scale of the earlier monsters. Neither of these kits were good sellers, because models of female characters don't appeal to ten-year-old boys. The Witch would stay in the Aurora catalog, but the Bride of Frankenstein departed early. Mold shop supervisor Henry Kolodkin remembered that the mold for the kit had never worked right, and the kits' poor sales didn't justify fixing it.

The final monster model, The Forgotten Prisoner (422), reverted to 1/8 scale, but continued the diorama approach. Bill Silverstein made a deal with Warren Magazines: Aurora would

Warren's Creepy magazine carried a story in its August 1970 issue that created a tale to explain the chained Forgotten Prisoner (422).

make the model if Warren would promote it. Silverstein thought kids were fascinated with skeletons, but this kit pushed the limits of good taste, just as Warren's magazines did. Although the Forgotten Prisoner was created out of thin air, *Creepy* (August 1970) carried a story to explain how the Prisoner came to be forgotten.

King Kong's Thronester (484 $400-500) attempted to unite two popular model genres: monsters and hot rods.

With monster madness at its height, Aurora began releasing another series of kits designed to combine the appeal of two popular models: monsters and cars. This resulted in the Monstermobiles—Universal movie monsters driving hot rods. Silverstein had advised against ruining the "purity" of the monster line, but the monster cars got the nod to go into production. Late in 1964 Frankenstein's Flivver and Dracula's Dragster (465-466) were released, followed in 1965 by Wolf Man's Wagon (458), Mummy's Chariot (459), King Kong's Thronester (484), and Godzilla's Go Cart (485). The collection did not gain much of a following, perhaps because the monster and car themes just did not mix. No auto modeler would have bought them for their automotive qualities.

When Aurora commenced its line of monster kits, art director Si Friedman went to Manhattan's largest commercial art company, Cooper Studios, to find an artist for the series. The assignment ended up on the desk of James Bama, a good friend of Aurora's hot rod artist Mort Künstler. Bama had won acclaim for his dramatic illustrations in men's magazines and would soon become famous for the covers he did for Bantam Books' *Doc Savage* paperbacks.

Bama enjoyed the thrill of recreating the excitement which he had experienced as a kid when he first saw the old horror movies. Working from Universal's publicity still photos, Bama transferred Boris Karloff, Bela Lugosi, and Lon Chaney, Jr. into colorful box art. Later some people complained that his illustrations on the Dracula and Wolf Man boxes did not match the poses of the models inside. This had happened simply because Aurora was in a hurry

to get these kits out; so sculptor Bill Lemon and artist Bama could not coordinate their labors. Thereafter Bama painted forthcoming kit art from photos of Lemon's model prototypes. He used both a photo of the prototype and a Universal still to paint Anthony Quinn as the Hunchback. Later Aurora removed Quinn's face from the Hunchback's box art without giving Bama an explanation.

Bama was also chosen to do the art work on those kits which Aurora wanted to relate to the Movie Monster line: Gigantic Frankenstein, the Guillotine, the two Customizing kits, the Addams Family Haunted House, and the Munsters. When the Monstermobiles came out, Bama objected that their cartoon-like style did not match his artistic talents. He painted the first four "under duress," considering a blood-drinking Dracula motorist "garbage." Friedman left photos of the last two monster dragsters, King Kong and Godzilla, on Bama's desk while he was out to lunch, hoping that Bama would relent and do them. He would not. Nor did he paint the box art for the last regular monster kit, the Forgotten Prisoner (that went to his friend Mort Künstler).

James Bama used his wife as the model for The Witch (483 $250-300). "It was her favorite modeling for me," he declared!

By this time Bama had soured on life as a commercial illustrator. He realized that both Aurora and Bantam's artist fee remained stuck at $300 to $600 a painting. Thus in the fall of 1968 he packed his family in a car and, departed for rural Wyoming. Since then he has become one of the country's most prolific and respected Western artists. His fine art sells for much more than $300 a copy.

In 1969 Aurora introduced a startling innovation in an attempt to rev-up interest in its aging monster lineup. Wholesalers visiting the Aurora parlor at the HIAA trade show were ushered into a darkened room by a hostess who would flip on an ultraviolet light and let out a scream. On display were six of the Movie Monsters with glow in the dark heads, hands, and detail parts.

This "Frightening Lightning" set culminated years of thinking. The idea of glow in the dark kits had been around for a long time, but phosphorescent plastic was just too expensive. Finally, Shikes located a dependable supply at a fairly reasonable price; so Aurora decided to include some glow parts in each kit. A full set of regular parts for each kit came in the box, along with assorted glow parts made from a separate mold. Kit builders had a choice of which parts they wanted to use.

Glow in the dark plastic gave the monsters a new lease on life.

The first six glow kits (449-454) came in standard "flat" rectangular boxes with a lightning bolt reading "Frightening Lightening" across the box illustration and a "Glows in the Dark" seal in one corner. New versions of Bama's original box art were painted with highlights to illustrate the phosphorescent properties of the models inside the packaging. Later in 1969 Aurora changed to "square box" packaging for the glow monsters. The only original monster that did not make the transition to the glow reissues was the Bride of Frankenstein.

During the height of monster madness, Aurora issued two spin-off kits from the popular monster TV shows to satisfy customers' demand for weird stuff. The Addams Family Haunted House (805) was the more popular of the two. The Munsters (804) created a cozy, humorous diorama scene of the Munster family at home before the fireplace. The model required so much assembly and intricate painting that most kids didn't stand a ghost of a chance of completing it. The kit was produced for such a short time that it did not find its way into any Aurora catalog.

To build on the success of the monsters, Silverstein came up with another idea: Madame Toussaud's Chamber of Horrors. Si Friedman approached London's famous Toussaud's Wax Museum and asked for the rights to do any of their figures. (Actually, Aurora intended to reproduce only the Chamber of Horrors torture devices.) Bernard Toussaud agreed to allow Aurora to use his museum as a source of information and to put the "Toussaud" name on Aurora's kit boxes. In return he asked only that the kits carry the message: "When in London visit Toussaud's." And he stipulated that Aurora must get his approval before issuing each Toussaud model.

The first kit proposed for the "Chamber of Horrors" series was the Guillotine (800). The design called for a working replica of a guillotine and a prisoner with his hands bound behind his back. By yanking a string the blade could be made to fall, knocking off the prisoner's head. Aurora asked Ray Meyers to carve the pattern, but he refused. "I didn't think that was the sort of thing kids should be monkeying around with." So Larry Ehling did the work.

Ray Haines loved this nifty working model, but other folks definitely did not. Because Aurora had issued the kit without first submitting it to Toussaud for approval, Toussaud sent Friedman a cease and desist order saying that Aurora had violated the terms of its agreement and demanding that Aurora stop using the Toussaud name. Thus only the first run of kits reads "Toussaud." Another blast

The glow kits included all the original parts to the model, plus a new tree with some duplicate parts in phosphorescent plastic. Frankenstein in a square box (449 $140-160).

New York City was the center of commercial art in America, and Aurora had access to some of the best illustrators in the country. The company's contact with the art community was Si Friedman, an artist himself who ran a graphic design company in Manhattan. Friedman knew the artists on a personal basis and easily made deals with them to supply Aurora with great art at low prices. Part of Friedman's success was his easy-going approach to assignments. He would give artists several assignments at one time and allow them to complete the work at their own pace. Of course, the pace of any commercial artist had to be pretty rapid.

Friedman would work with Silverstein, George Burt at the print plant, and Shikes to pull the art, graphics, and kit package together into a finished product. The idea was to create a uniform "look" to a series of kits so that customers in stores would instantly recognize individual models as parts of a familiar product line.

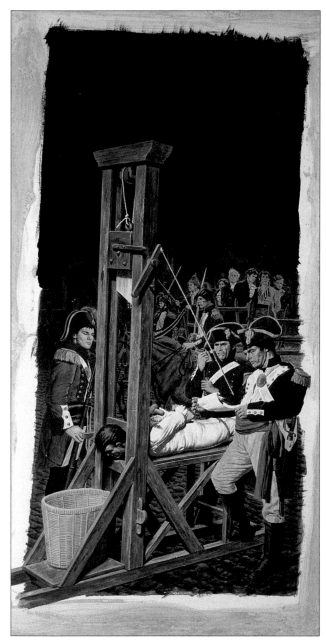

Bama used an old publicity still photo from the 1920s silent movie *Orphans of the Storm* as the reference for La Guillotine (800).

of criticism came from the country's top television comedy show Rowan and Martin's *Laugh In*. They ridiculed the Guillotine by awarding it the "Flying Fickle Finger of Fate" for bad taste.

Following these troubles, the rest of the Chamber of Horrors series got the axe. Next in line were the Hanging Tree, the Rack, the Electric Chair, and the Gallows. (Patterns for the first three were completed) One wonders what Rowan and Martin would have said about the Rack ripping out the prisoner's arms.

Much of the terrific success of Aurora's monster model series—indeed all its kits—can be attributed to the fabulous illustrations that went on the lids of Aurora kit boxes.

Patterns for The Hanging Tree and The Rack were sculpted, but negative reaction to the Guillotine consigned these next two Chamber of Horrors models to Aurora's storage vault.

The Molding Room

For twenty years Henry Kolodkin's office was the Aurora molding shop. Here, in an atmosphere of noise and heat, the real business of Aurora was done. Dozens of machines full of 500 degree hot plastic made the shop as warm as a bakery. The water used to cool the machines was piped on to the roof where lawn sprinklers sprayed it around in a largely fruitless attempt to cool the building off.

Henry Kolodkin removes a tree of parts for the B-58 Hustler (375 $30-100) from an injection molding machine.

The noise of machines closing, plunging molten plastic into the molds, and springing open was deafening. "You couldn't hear yourself think," said Kolodkin—who habitually spoke in a loud voice even outside of the shop. Each of the forty machines had a distinct operating sound, and a skilled worker could tell by the sound whether a machine was running properly or not.

Aurora did all its packaging right at the machines in an attempt to maximize efficiency since every split second multiplied labor and machine costs. Each machine operator had a table next to him stacked with supplies. While the machine stood in its closed phase, the operator would put an instruction sheet, decal sheet, stand, and clear canopy or other auxiliary part into an open kit box. When the mold sprang open, he would quickly remove the shot, clip the runners holding the parts at specific points, drop the parts in the box, put on the box top, and place the box in a packing carton. A normal cycle for this process might be about 25 seconds—150 shots an hour. Once every two hours the operator got a break.

Kolodkin and a crew of inspectors saw to it that the shots were coming out of the machines perfectly. If a part came out improperly molded, or if "flash" from seepage between the mold halves appeared, the machine would be shut down immediately and the problem corrected. Giammarino visited the molding room frequently. "He was a stickler for detail" said Kolodkin. "If he'd walk through the plant and see a drop of oil, he'd raise holy hell, and have that machine shut down immediately."

Because of this vigilance, the only time a machine was shut down came when it was repaired by Aurora's staff of machinists or when a run of kits had been completed and the mold had to be changed.

Switching molds required from 40 minutes to two hours. A chain hoist would be rolled in, the mold clamped shut, lifted out, and then wheeled to its designated spot on the racks lining the walls. Molds for popular kits had convenient parking spots, while less used molds went to some far corner of the plant. If a mold being placed in the machine was a brand new one from Ace or Ferriot, it would be thoroughly tested and adjusted before production began. Sometimes a mold had to go through a dozen or more trials before it worked properly.

Each afternoon at four o'clock, the night foreman came in, and Kolodkin would go over the evening's production schedule with him. By seven Kolodkin had visited Giammarino for a final chat about the day's work, and he could go home. Often at two a.m., his telephone would ring, and the third shift foreman would say, "An ejector pin has broken on the number three machine, do you want to come down and fix it?" The drive to the plant was about ten minutes, and he made it frequently in the dark.

Chapter 5
Modeling's Golden Age

A golden age of plastic modeling emerged in the 1960s. More hobbyists were building models than ever before, and hobby companies offered them an ever-widening variety of high quality kits. Nationwide model sales had totaled $95 million in 1957, and some people worried that the market had become saturated, but by 1967 sales had more than doubled to $224 million. The typical hobby shop made about a third of its profits from kits—far ahead of slot cars, craft supplies, and model railroading which followed in that order.

Model cars emerged as the kits with the greatest appeal to American boys. Companies like AMT, MPC, Monogram, and Revell were bringing out models of the latest Detroit cars and European sports cars, as well as drag racers and custom show cars. They scaled their kits to 1/24 or 1/25 scale and charged a pretty high price for their kits so that lots of parts and fine details could be built in.

The 1/32 scale MG-TD (590 $50-60) had bright blue plastic and nifty vinyl tires, but like most model cars in the 1950s, it lacked clear and chrome parts.

Model cars posed a challenge for Aurora, which had built its reputation on aircraft, ship, and figure models. Its initial offerings in the automotive realm had been all over the field: ex-Best Indy cars, ex-Advance sports cars, the antique Old Timers, and the offbeat motorized toy-model cars. The boys at Aurora headquarters considered their options and decided to follow-up the path begun by the Advance cars. Aurora's next cars would be smaller 1/32 scale kits and would be simpler and less expensive. This fit the company's basic philosophy.

Thus in 1960 Aurora issued its first original car kits: the Austin-Healey 3000, Mercedes-Benz 300SL, and Triumph TR-3 (516-518). These models were designed by HMS and included the same features as the Advance cars. They had no chrome or glass parts, but did have black vinyl tires in wheels that snapped on the chassis. In subsequent issues HMS improved the 1/32 sports car line by adding clear plastic windshields, chrome-plated parts, and a standard wheel attachment on a plastic axle. By 1965 new car kits were being issued without vinyl tires. The plastic tire parts that replaced them didn't look as authentic, but were cheaper to manufacture. These .49, .69, and .79 cent models were simple, easy to assemble kits aimed at youngster modelers.

The distinctive lines of the Porsche Carrera (539 $40-45) made it a favorite for model makers.

The '65 Mustang (665 $30-40) was one of Aurora's best selling kits.

One project in 1/32 scale, however, broke new ground in detail and quality: the American La France Pumper (599) fire engine. Ray Haines traveled to the American La France plant in Elmira, New York to get photographs and blueprints of the pumper. The kit that emerged from this project had 219 parts molded in seven different colors, with real rubber tires. The problem was that its complexity made it difficult to manufacture and sales were never very high. It did have a four year run in the mid-60s, and it eventually returned as the California Fire Truck (599) in 1971-1972.

Performance auto parts designer George Hurst supplied technical advice to Aurora to help them make two of his cars, the Hemi Under Glass (680) funny car and the Baja Boot (681) off-road racer. These were two of the more detailed 1/32 scale cars in Aurora's line. Hurst made a "handshake agreement" with Haines for use of his name and company logo on the cars. After Haines's departure from Aurora, Hurst withdrew his consent, and Aurora removed both cars from the catalog.

Because 1/32 scale cars were inexpensive for Aurora to develop, it was possible to make models of some exotic cars like the Chevy Monza SS (675 $50-60) even if they weren't likely to attract high-volume sales.

Kids like to smash things up. So why not offer them a model of a Demolition Derby (673 $80-90) car?

To recycle some of its older models, Aurora turned them into custom jobs, with lots of optional add-ons. Hidden under the Astrovette's (547 $30-35) accessories is the '61 Corvette (519).

To go with its sports cars, Aurora developed a line of street rods, and these little kits came to epitomize Aurora's car kit line. They caught the 1960s surf 'n rock spirit of the Beach Boys and songs and B movies celebrating Jags, Vettes, and Deuce Coupes. Boys who were too young to drive real cars could join in the fun by building Aurora's hot rods and turning their imaginations loose.

Aurora's first three hot rods. The Ram Rod (509 $45-55) was Aurora's version of Joe Cruces' Deuce Coupe that appeared on the cover of the Beach Boys' album.

The Street Rods were classic 1920s and 1930s cars, customized into wild California hot rods. Most of the hot rods were based on actual show cars, but Aurora didn't acknowledge this and made up new nicknames for each car. The price of these kits was held down by keeping detail simple and omitting chrome-plated parts. The molds were cheaper beryllium copper, rather than more expensive machine-tooled steel. At .49 to .79 cents, you couldn't take them very seriously. They were just fun, and priced about right for a kid's weekly allowance. Aurora eventually had twenty-two cool rods in its catalog.

Automobile box art was the realm of several artists, but most notably of Mort Künstler, a well-known freelance magazine illustrator. Born in Brooklyn in 1931, Künstler began sketching as a sickly, shut-in child, but when he grew up to be a robust star basketball player, he still kept his love of drawing. He studied at the Pratt Institute and became friends in New York art circles with James Bama. Aurora first hired Künstler to create new covers for several of the old World War II aircraft kits.

The Show Trailer (530 $20-25) attached to the Chevy Pick-Up (555 $45-55) and could haul the 32 Skid Doo (554 $45-55)—or any of the hot rods.

Add bright plastic and snazzy decals to a radical hot rod creation, mix well, and you have "Road Raider" (551 $50-60).

The Ford Woodie (558 $50-60) makes a perfect surf wagon.

For Aurora, Künstler did aircraft, the boat Wheeler Cruiser (444) and some figure kit boxes, such as the knights, the Men from U.N.C.L.E (411, 412), and John Kennedy (851). However, Künstler's most charming art appears on the boxes for the street rod cars. These have a *Saturday Evening Post* cover flavor, with each painting telling a story, sometimes with a humorous twist. Si Friedman worked with Künstler on the hot rods to make sure that each car stood out as a distinct customized hot rod, yet each car also could be seen as one of a set. Friedman had selected Künstler for this work since Künstler had a reputation for being good with complex compositions and perspective. The hot rods posed a special problem because Aurora gave him a photo of the cars taken from a three-quarter elevation. Thus Künstler had to distort the perspective to paint a scene for the box top.

Although Künstler enjoyed working for Aurora and other clients who commissioned illustrations, he aspired to improve the quality of his art and increase the prices his art would command. In 1965, *National Geographic* hired him to do some paintings about the founding of America's oldest city, St. Augustine, Florida. "It was such fun. It changed my whole viewpoint." Künstler decided to become a painter of moments from history. Today he is renowned as the foremost illustrator of the Civil War.

Künstler approached doing an illustration by first gathering reference photographs and then building a model of the plane. "If you are looking at a real plane," he explained, "it looms at you when you are close to it. There are no photos that give you that dramatic angle. So I'd take the model and hold it close and shut one eye. I would hold the model at dramatic angles. Then I got a camera that could take close-up pictures—so close that it would fuzz out. But I would get the general idea, and then I knew the details to fill in." He would use a projector to throw an image of his plane or car on art board, then pencil in the outline to get the angles right. Finally he would go to work with his paintbrush creating an artistic rendition of his subject.

Mort Künstler's box illustration for Old Ironsides (663)

Künstler used a photo of the family to help compose his cover for The Charger (536). The boy waving is his son, the two girls are his daughters, the young lady in pink is the artist's wife, and the model for the cop was the artist himself.

In 1961 a car debuted at the New York Auto Show that electrified the automotive world: the Jaguar XKE. Low, sleek, aerodynamic, and expensive, the XKE redefined the sports car profile. Aurora's Ray Haines stood in the crowd. "I saw the XKE at the Auto Show in New York, and I took pictures of it. I brought it back to Joe, and I said, 'Joe, we've got to make this car, but 1/32 is too small to get the features in.'" Giammarino's favorite car in the Aurora lineup was the Jaguar XK-120; so he liked the idea of adding another stylish Jaguar into Aurora's offerings.

Thus Aurora moved up to 1/25 scale cars. The XKE model had more than 100 parts, a fully-detailed interior, and the front end lifted up to reveal the engine. However, the feature that really captured the attention of the modeling world was the wire wheels. HMS and Ace had perfected the first one-piece wire wheels that really looked like spoked wheels. They created the illusion by molding the wheel in black plastic with very deep recesses between each spoke and then plating just the spoke surfaces—not the black plastic recesses behind the spokes. Rubberized vinyl tires with white plastic inserts to make whitewalls completed the package.

When this model was reissued by Monogram in 1990, almost thirty years after its original release, a reviewer in *Scale Auto Enthusiast* declared: "It was way ahead of its time in terms of detail back then and can certainly hold its own today."

By placing the Moody Monster (537 $60-70) at an angle, Künstler was able to bend the horizon line a little and stage a scene that seemed to have correct perspective.

A sports car, like the Maserati 3500 (564 $20-30) would not go out of style as quickly as one of Detroit's annual domestic models. Thus it could stay in the Aurora catalog for years.

The Porsche 904 (561 $20-30) was one of the most popular grand prix racers of the 1960s.

Aurora went on to complete a set of nine 1/25 scale sports and GT cars. These models were based upon the latest prototype cars which raced over such popular road circuits as Sebring and Watkins Glenn. They were molded in two colors of plastic and had opening doors, hoods, and trunks. At $1.49, and later, $1.98, these kits were aimed at quite a different audience than the .49 and .69 cent cars. Jim Keeler, then a R & D man for Revell, recalled wondering what the heck had gotten into Aurora.

The '34 Fords, Stock and Custom Rod (569 $130-150), were among the best car models Aurora ever made.

In 1964 Aurora made a dramatic bid to further strengthen its presence in model cars: it entered into an agreement to merge with AMT, the model car specialists of Troy, Michigan. Under the proposed agreement, AMT would exchange its company stock for a one-quarter interest in Aurora. AMT would become a subsidiary of Au-rora. The success of Aurora's slot cars and monster models had already made it the largest hobby company in America. The acquisition of AMT would greatly expand Aurora's offerings in the most popular line of kits. However, the deal did not come off. On August 24, 1964 Aurora announced, without elaboration, that the merger had been abandoned.

Aurora's press release did not disclose that Lew Glaser of Revell had filed a complaint with the Federal Trade Commission charging that the merger would give Aurora too much control over the hobby industry. Aurora responded that it was a small company compared to other toy companies, but the FTC said that they were a hobby company, not a toy company. Aurora's claim to be the "World's Largest Manufacturer of Hobby Products" had been held against it.

In fact, Aurora had grown to be a relatively large company even when compared with the toy companies. In 1969, *Business Week* would rank Aurora seventh in sales behind Mattel, Milton Bradley, Ideal, Tonka, Hasbro, and Remco. Aurora would have been eighth if Fisher-Price, a part of Quaker Oats, had also been counted.

Although Aurora did not get into show cars the way Monogram and Revell did, it created a couple of models in the 1960s that have become collector classics. One is the Super Spy Car (585).

The Super Spy Car (585 $120-140) didn't have much appeal to serious car modelers.

Aurora took its 1/25 scale Aston Martin DB4 (562) and reproduced it as the car James Bond drove in the movies. However, Aurora could not use Bond's name on the kit since Airfix held the Bond license. Thus the "Spy Car" label. It had many operating features, the most dramatic being the spring-operated ejection seat that sent the passenger through the roof. Serious car model buffs judged the original Aston Martin DB4 superior to the spy version.

The most famous Aurora car today is Carl Casper's Undertaker Dragster (570). The real show dragster won top honors at the National Hot Rod Association's 1963 championships. It featured a padded, coffin-style interior and chrome skull hood ornament. Casper took the car to

Willow Grove for the HMS model builders to copy and then put it in Aurora's exhibition parlor at the HIAA annual convention in Chicago. The car model, with its monster-kit base, two sinister figures, and tombstone, looked right at home during 1964's monster fad.

Abe Shikes, Big Frankie (470 $1200-1500), and Carl Casper pose in the Aurora parking lot with Casper's Undertaker dragster.

Because plastic cement is an essential ingredient in the model building process, Aurora set up its own glue production facility shortly after it started making model airplanes. The chemicals for the cement were mixed in a huge tank, then piped into an automatic tube-filling machine. Working in the cement manufacturing room was one of the least popular jobs at the plant because of the overwhelming toxic fumes. "I think some people are dead today because of it," Frank Carver later mused.

In the early 1960s, some American youngsters sought out those fumes for their narcotic effect. They would empty a tube of plastic cement into a paper bag and then inhale the vapors from the solvents and quick-drying agents. Some kids dropped dead. Suddenly "airplane

"Casper's Ghost," by Carl Casper, never made it into mass production.

glue" sniffing became a nation-wide crisis. Local and state governments were prompted to consider bans on the sale of model cement. If enacted, such laws would kill the model kit industry.

Faced with this catastrophe, the HIAA set up a committee to combat restrictive legislation and to find a way of making a safe cement. The industry produced an educational film to inform the public that plastic cement was harmless when used properly. The HIAA also urged hobby retailers to sell tubes of cement only to customers who also purchased a model kit. Within a few years Testor Corporation discovered that simply adding a common chemical called oil of mustard to the cement made inhaling the fumes so unpleasant no one would do it. This solved the problem for the hobby industry; however some self-destructive youngsters simply moved on to other products that produced fumes that could be inhaled.

Low Flying Aircraft

During the 1960s Aurora's aircraft models failed to keep up with the competition's improvements in accuracy and detail. In the 1950s, Aurora's airplane kits had been only a little simpler than Revell's and Monogram's, but in the sixties the quality gap widened.

This painting for the Northrop P-61 (392) was one of the last—and one of the best—Jo Kotula executed for Aurora.

In fact, Aurora took a step backward in May of 1963 when it purchased Comet Model Company's plastic kits and flying aircraft. Twenty-six Comet aircraft went into the Aurora line. Several of the kits duplicated existing Aurora models. Some of these were given their own packaging and catalog numbers, but others went into the same boxes as their Aurora counterparts. Today the Comet models can be distinguished by their characteristic teardrop display stands.

While the Comet models varied in quality, most were distinctly 1950s in their standards (production manager Frank Carver remembered the molds as "dogs"). By the late 1960s the Comet kits began to disappear from Aurora's sales list.

Aurora created only a few new aircraft kits during the 1960s. The .49 cent aircraft kit series came to an end with the introduction in 1962 of the F7U Cutlass, F4F Wildcat, and P-38 Lightning (496, 497, 498). Only six kits of up-to-date 1960s era warplanes were released: the F4U Phantom (391, 394, 367), the Northrop N-156 (140), the XB-70 (370), the F-111 (368, 369), the C-141 Starlifter (376), and the A-7 Corsair (395).

In 1/48 scale, the F4U Phantom II turned out to be a large model by fighter aircraft standards; so Aurora sent its mold to Fellows Gear in Bellows Falls, Vermont to see what size injection molding machine it would require. Fellows, which supplied Aurora with most of its machines, was pleased to demonstrate that one of its medium size machines could handle the job. Interestingly, although the Phantom was a poor model based on inaccurate information about the plane's prototype, it outsold the more accurate A-7 two to one.

The Air Force test plane version of the McDonnell Phantom was labeled the F-110 (391 $60-70), but it went into production as the Navy's F-4.

Of the new military kits, the F-111 had the most detail. It featured retractable landing gear and operating swing wings, just like the real aircraft.

Some new aircraft models made it as far as the pattern stage before being canceled. Among the patterns which went into the storage vault were an F4U Corsair, F-14 Tomcat, an A-6 Intruder, and two versions of proposed Boeing SST airliners.

Roy Grinnell enjoyed doing the few box covers he painted for Aurora. The color sketch he submitted for approval before doing the final art looks very much like the illustration that went on the box. The Navy's version of the TFX F-111 (369 $30-40).

Merit of Great Britain made pirate copies of Aurora's first six World War I fighters. The molds later were sold to Artiplast of Italy and today are still being used by SMER of the Czech Republic.

In May of 1961 Aurora received a "to whom it may concern" letter from two California freelance illustrators asking if Aurora would like to hire them to paint box covers. The artists were John Steel and Jack Leynnwood, and they were the mainstays of Revell's model box art team. Joe Giammarino immediately recognized Steel as the creator of Revell's popular "Picture Fleet" series of ship model kits. Bill Silverstein, who was in the midst of a program to "refresh" the look of Aurora's packaging, jumped at the chance to hire Steel and Leynnwood.

Aurora made a pattern for an F-14 Tomcat, but never produced it in kit form. *Photo by Rick DeMeis.*

Steel flew to New York to meet with Giammarino and Silverstein, who found Steel to be "very professional and an excellent artist—outstanding." Steel had been born in New York, but his actress mother took him to Europe, where he grew up and became fluent in French and Spanish. Later he even took a turn acting and singing on Broadway, but found his true home in the Marine Crops during World War II and Korea. He fought and suffered several wounds in some savage battles, yet he also took time to develop his artistic skills, doing sketches of the combat. After World War II he studied under the G. I. Bill at the Art Center in Los Angeles and became the cover artist for *Skin Diver* magazine, as well as painting for several other clients including Walt Disney.

Jo Kotula's painting for the USS *Bennion* (704 $25-55) was shaded toward the orange end of the spectrum, while John Steel's replacement art trends into the blue range. Kotula greatly admired Steel's work.

Silverstein took Steel out to dinner on their first meeting and explained that he wanted to give Aurora's packaging graphics a cleaner, more modern look with the focus on a "dramatic portrayal" of the model that would give "pop" to the product. Steel's awesome paintings of hulking steel warships plowing through deep ocean waves were just the thing Silverstein had in mind. Steel's rugged treatment also worked for armored fighting vehicles and aircraft. However, Steel loved doing wildlife art, and his softer images for Aurora's animal models worked very well.

Steel's painting of the Sherman tank (329) has had a panel added to its top and the background painted over in pink in order to make it fit the square format adopted in 1970. The painting had originally been rectangular.

Steel's associate Jack Leynnwood also came out of the Renaissance man mold. He had been born near Los Angeles in 1921 and as a youngster had filled some bit parts in *Our Gang* movie shorts. A child prodigy musician, he toured Depression-era America playing saxophone to help pay the bills for his family. During World War II he became an Army Air Corps fighter pilot, but also painted some poster art. After the war he studied at the Los Angeles Art Center and then went to work as a commercial artist. Northrup Corporation hired him to paint illustrations of its missiles and aircraft. One of the artists he admired was Jo Kotula, whom he called "a colorist with no inhibitions." By the late 1950s Revell had made Leynnwood one of its chief artists, and that led him to Aurora.

Silverstein thought that Leynnwood's smooth, tight painting style would work well with jet airplanes and sports cars. He would send Leynnwood requests for two or three preliminary "comprehensive sketches" of a proposed box cover, then Silverstein would instruct Leynnwood on what he wanted for a "final." They coordinated efforts by mail or telephone, and the arrangement worked well. At the end of their first year of collaboration, Silverstein wrote Leynnwood: "I think every piece of art work you have sent us is outstanding and will certainly enhance the appearance of our company's line."

Unfortunately for Aurora, Leynnwood became so involved with his commissions from Revell that after the mid-1960s he had to quit accepting requests from Aurora for additional art.

Jack Leynnwood's KC-135 (143) seems about to zoom right off the canvas.

Leynnwood painted an olive drab P-47 (81 $25-35) in his preliminary composition, then switched to bright metal in the final art.

Bill Silverstein instructed Leynnwood to create a painting that emphasized the engine pods of the Boeing 727 (353). So Leynnwood put the 727 in a night scene with the strongest light falling on the tail section.

During the 1960s, helicopters received top billing among the military aircraft. This was the time of the Vietnam War—a "chopper" war. Aurora kept the old Sikorsky S-55 Windmill (503) and the Piasecki H-21 Workhorse (504) in the catalog, but the rest of the ex-Helicopters for Industry kits were replaced. The HC-1B Chinook, CH-54A Sky Crane, and Sikorsky HH-3 Jolly Green Giant (350, 499, 505) were big transports. The Bell UH-1B Cobra, Bell AH-1G Assault 'Copter, and Lockheed AH-56A Cheyenne (500, 501, 502) were nimble gunships. The two Bell attack helicopters were among the most popular kits in Aurora's line, outselling all of the aircraft models in most years.

Roy Grinnell's color sketch for the Bell UH-1B (500) was almost good enough to go right on the box.

Civilian airplanes never had as much appeal for boys as their death-dealing warplane counterparts. However, most model companies put at least a few civilian airliners and light planes into their catalogs. Aurora's first airliner kit, the famous Boeing 707 (381), came out in 1958 and proved to be a good seller.

The 707 set the standard for airliner kits that followed. It had good basic outline and part fit, but there was no glass for the passenger windows, no wheel wells for the landing gear, and little surface detail on the fuselage and wings. Despite these shortcomings, those modelers who built civilian jetliners were happy to have the 707 and its forthcoming sister ships in plastic.

Aurora built up a fleet of passenger aircraft, although none of the new jetliners enjoyed the popularity of the 707. From Boeing, it offered the 707, 727, 737, 747, and 720B. Aurora built the 727 at the request of Boeing, which offered technical help developing the kit. Douglas was represented by the DC-8, DC-9, and DC-10. Convair had the 880 and 990. All the basic models were issued as an array of different kits simply by printing a variety of decal sheets with markings for different airlines.

Civilian light aircraft was represented in the Aurora catalog only by five simple kits purchased from Comet. Flying enthusiasts bemoaned the model companies' refusal to produce kits of the aircraft that could be seen parked at every local airport. The model companies, however, knew that civilian planes just didn't sell.

The light aircraft pilot on Aurora's staff, Ray Haines, had argued for years that the company should offer a small line of civilian aircraft. Aurora completed the patterns and decals for a set of six, but then the project was put on hold. Finally, Shikes agreed to complete the series, but only if the tooling was done by a company in Portugal that promised to do the job for half the price Ace charged. Haines and Shikes flew to Portugal with the patterns and hand delivered them to the tool company.

The Boeing 707 (380 $45-55) set the standard for its airliner series: fairly simple models that could be issued with different decals for various air lines.

The Boeing 737 (359 $35-45).

When the completed molds arrived at West Hempstead, Joe Giammarino inspected them and declared: "Throw them out." Shikes admitted, "I guess we made a mistake." The molds for a Lear Jet and a Cessna Skylane were scrapped, but the Jet Commander (85), Cessna Skymaster (279), Piper Cherokee (281), and Piper Aztec (282) went into production. Despite their low quality and outdated decals, the kit filled a void for some airplane enthusiasts.

Roy Grinnell placed the United insignia in the same position used on Aurora's prototype in his initial color sketch; then for the final box art he moved the insignia higher on the plane's fuselage.

In the mid-1960s some of Aurora's World War II aircraft were revised to capitalize on the TV series *Twelve O' Clock High*. Three of the small B-17 (491) models were given a common landscape base with three arms to hold the bombers in formation to create a diorama kit (352). Seven other models were put in boxes with a *Twelve O' Clock High* identification band. The B-25 and B-26 got new decals to represent "battle damage." All these kits were molded in olive plastic, not their usual assortment of colors.

Other aircraft that made a comeback in the late 1960s under a new guise were Aurora's gas powered U-control flying airplanes. Two ex-Comet planes, the P-51 and P-40 (377, 378) also joined the series. These were converted to static display models by removing the engines and replacing the oversized propellers with new props of the proper scale. Since these planes had originally been assembled at the factory and were too big to be cemented together, they were redesigned to be assembled with metal screws.

These "Screwdriver" kits made big, but not very accurate models, and Aurora pitched them to youngsters. They outsold all of Aurora's aircraft in 1968 (except for the helicopters) because the big chain stores sold them as "toys." However, Giammarino recalled that serious modelers like singer Mel Torme said that with modifications and scratch building the screwdriver models could be turned into good display pieces.

In 1966 Aurora ran a second "Idea Contest" asking model makers to send in their suggestions for a new kit and offering a $1,000 prize to the winner. This contest was partly designed to find out what kinds of models were in demand, but other objectives were more subtle. Aurora wanted to know the ages of model builders, the stores where they purchased kits, and their favorite TV shows and magazines. When the entries came in, the staff were amused to discover that many preteen boys listed *Playboy* as their favorite magazine!

The B-25 (397 $180-200) first appeared in 1960 as a flying model, then was retooled to become a screwdriver-assembly model in 1969.

The familiar black Focke Wulf 190 (344 $100-120) became olive in the 12 O'Clock High set.

Some Very Good Ships

During the 1960s, Aurora's modern warship line did very well. HMS's ship creations received good reviews from serious modelers and also sold well. If one discounts the toy-like *Halford* and *Nautilus* kits of the early 1950s, Aurora's ship line began in 1957 with the *Forrestal/Saratoga* kits (701-702). These were made from the same mold, with different decals. Along with these large aircraft carriers came the battleship *Iowa* (705), the heavy cruiser *St. Paul* (703), and the destroyer *Bennion* (704). Aurora had intended to name the destroyer *The Sullivans* to please Mattie Sullivan, but Revell got its *Sullivans* kit out first. Revell's ship models were usually a little larger, a little more detailed and a little more expensive than their Aurora counterparts. For example, Aurora's *Iowa* measured seventeen inches, compared to twenty inches for Revell's *Missouri.* Yet the deck part of Aurora's battleship nestled down snugly inside the hull—an improvement over Revell's parts arrangement, which placed the deck part on top of the hull, leaving a visible seam.

In 1959 Aurora issued a model of the German raider *Atlantis* (710), a converted merchant ship that sank Allied ships during World War II. The model, like the real ship, appeared to be an innocent freighter, but could be converted into a warship by folding down the sides and removing some parts to reveal guns, torpedo tubes, and a scout aircraft. It was a nice, big kit, but expensive at $2.49, and it took up a lot of precious space on store shelves. Thus it went out of the catalog in 1964 (but returned in Holland and England in 1973).

The German Raider *Atlantis* (710 150-200) is one of the toughest Aurora ships to find today.

The USS Iowa (705 $20-70) kit built into a nice, neat desk model.

Jo Kotula gave the Japanese battleship *Yamato* (713) a blazing battle action box illustration.

Aurora's most expensive ship kit, the nuclear powered carrier *Enterprise* (720) came out in 1961. This mammoth, three foot long model could be motorized and sailed on ponds. The flight deck lifted off to provide access to the motor inside. At $11.95 it was Aurora's most expensive kit, and its size made it hard to manufacture. The parts were scattered among four molds, and the large, one-piece hull took up a mold by itself. Still, it built up into an impressive model, and good sales kept it in the catalog right to the end. Aurora's aircraft carriers were one of the company's best selling lines.

Aurora's Wolfpack U-boat (716) came out in 1962 and remained the only U-boat on the market for the rest of the decade. The model was a copy of the U-505 which had been captured during the war and put on permanent exhibit in Chicago. The usually fussy grownup modelers praised this kit for its accuracy and detail, and when Aurora brought out the Japanese I-19 sub (728) in 1970, *Scale Modeler* magazine called it "the finest submarine kit ever introduced to the market." And it cost only $1.50.

For a few years Aurora did not release any new models of wooden ships, but then Revell scored impressive sales with its big model of the *Cutty Sark*. Although complicated models of sailing ships have only a small following among modelers, Aurora decided to create four ship kits that would build models over two feet long. Because tooling costs totaled $325,000, they had to be retailed at $5.95.

The nuclear-powered carried *Enterprise* (720 $125-150) was one of Aurora's most elaborate models. Today you can still find reissues on store shelves under the Revell label.

Boeing Company asked Aurora to do a model of its *Tucumcari* Hydrofoil gunboat (727) and supplied Aurora with blueprints. Boeing was probably trying to stir-up public interest in this prototype craft which it wanted to sell to the Navy. This kit was around only for a few years and because of its short run is rare today.

Over the years Aurora had produced a number of interesting sailing ships. The original *Black Falcon* pirate ship, *Viking Ship*, and *Chinese Junk* had been quite simple. The first authentic model of a real sailing ship was the Nova Scotian racing schooner *Bluenose* (431). Proud Canadians had immortalized it on the Canadian dime, and they furnished Aurora with drawings from which the model was made. 1959 saw the release of the *Cutty Sark* and *Privateer Corsair* (432, 433). These, like the *Bluenose*, had finely detailed parts, molding in three colors, and a reasonable price.

The series led off with the *Wanderer Whaler*, followed by the Civil War USS *Hartford*, the clipper ship *Sea Witch*, and John Paul Jones's flagship the *Bon Homme Richard* (440-443). The mold for each of these models was cast beryllium copper to capture the texture of wood. They were, in the words of molding foreman Henry Kolodkin, "bitches to mold. All copper and a million pieces!"

During 1967 and 1968 Aurora closed out its sailing ship line, in smaller scale, with the English warship *Sovereign of the Seas*, the Roman Bireme, and the USS *Constitution* (434-436). The Bireme was a cleverly patterned little vessel, but the *Sovereign of the Seas* and *Constitution* were very ordinary models.

The Viking Ship (320 $25-65) was a toy-like model, yet it has had enduring popularity. Aurora brought it out under three different box illustrations. Merit copied the mold and released its own issue. Today you can buy a SMER kit that is produced from the Merit tooling.

Aurora's model of the German battleship *Bismarck* (715 $35-60).

The old *Chinese Junk* model received a make-over in 1968 to capitalize on the public's interest in the naval war raging on Vietnam's rivers. A battery of machine guns were added to the kit's parts, and John Steel painted a new box cover depicting an explosive night combat. The Armed Command Junk (437) stayed on the store shelves for only two years, and then the Junk mold was retired for good.

Another nautical subject, the *Wheeler* Cruiser (444) was Ray Haines' pet project. He developed the pattern for this 65' fishing boat and a companion Chris Craft 38' Sport Fisherman in the early 1960s. However, the rest of the staff at Aurora thought that pleasure boats, like civilian aircraft, would not have a wide appeal. Finally, Ray talked Joe into going ahead with the *Wheeler*. It came out in 1967 and stayed around only until 1970.

If the Wheeler Cruiser was a good product with limited appeal, the American Cup Sailboat (724) was a bad product with virtually no appeal. Mattie Sullivan, always on the lookout for new products to sell, introduced an independent toy designer to Abe Shikes. He had a sailboat that could be remote controlled by a line played out from a reel and attached to a steering mechanism. Shikes wanted to give it a try because it might sell big in the toy market. Giammarino wanted to stick to hobbies and tried to talk Shikes out of it, but Shikes got his way. The sailboat's hull was manufactured of expanded styrene by Sullivan's company, while Aurora molded the deck and cabin. Trying to bring all the components together at the plant proved to be a disaster. The women on the assembly line tore their hair out over the complex control reel; so it was decided to sell the boat without the controls. A control pack was sold separately. Only a few thousand ever made it to stores.

The Wheeler Cruiser (444 $70-80) was a great model, but just didn't sell.

Tanks and other military vehicles rivaled automobiles as the most rapidly growing kit category of the 1960s. Interest in army vehicles had collapsed in the late 1950s, but by 1964 military armor was back in demand, partly spurred by the first imports of these kits from Japanese companies.

The Japanese first became interested in plastic modeling in the decade of the 1950s and had gained notice in the United States for their practice of making pirated copies of American models. Aurora was only one of a half-dozen

US firms to be knocked off. This led to protests by American corporate heads, as well as some interest in increasing exports of American-made kits to Japan.

For their part, the Japanese began exporting some new, original, and very high quality kits to the US. These new armor models were better than anything ever produced in the United States. The Japanese kits were high priced and hard to find, but the standards of detail went up a notch with their arrival in the market place.

Aurora issued the America Cup sailboat as a toy, but poor sales led to its quick demise.

Aurora responded by bringing back all its discontinued military kits (except the LaCrosse missile). Beginning in 1964 Aurora added seven more new armor kits to its line, including the first Japanese tank made by any company, the Japanese Medium Tank (313). Another unusual subject was the revolutionary Swedish S Tank (316). A writer for *Scale Modeler* (March 1966) declared: "If you're an armor enthusiast, you can't go far wrong if you invest in Aurora's 1/48 scale line of combat tanks and self-propelled gun kits. They are on the whole accurate, reasonable in price, and, even more important, readily available."

The fact that Aurora's tanks were 1/48 scale set them apart from the offerings of most other companies, which settled on 1/35 as the appropriate scale for armor. This shut Aurora out of the market among hard-core military vehicle modelers, but most of them were not Aurora fans anyway. The tanks would continue to occupy a secure niche in Aurora's catalog right to the company's end.

In 1964 Aurora began selling the Austrian company Roco's HO scale toy tanks packaged on blistercards. They inspired two of Aurora's best selling kits: Rat Patrol (340) and Anzio Beach (339). Some of the Roco tanks were incorporated into these sets, which also included landscape pieces such as trees, sand dunes, and pillboxes. They were essentially toy soldier sets and came out when Aurora was trying to increase sales volume by branching out from hobbies into toys.

U.S. ARMY HOWITZER ★ M109 TANK
AURORA PLASTICS CORP. • ALL PLASTIC ASSEMBLY KIT

John Steel's art for the Howitzer Tank (314 $25-30) captures the massive size of the steel monster.

Chapter 6
Stars of TV and Comic Books

Aurora won its enduring place in American popular culture with the figure kits it issued in the 1960s. At the time, Aurora simply dominated the market in figure kits. No other company came close in sales, or approached the quality of Aurora's products. After the success of the monsters, Aurora began looking for something else to do. Because of his monster models, Bill Silverstein was riding high at Aurora—indeed, in the industry—and he had some ideas.

Silverstein, the former minor league baseball player, was a sports fanatic. Seeing a nation of sports fans as a potentially huge market, he developed the Great Moments in Sports series. Diorama models were created with famous sports heroes captured in "stop action" at the moment of their greatest triumph. A model of Jack Dempsey was a natural for the men at Aurora, who knew him from his New York restaurant where he entertained guests with magic tricks. Silverstein paid him $10,000 to come to the HIAA show in Chicago and introduce the model of his classic

1923 battle with Luis Firpo. Aurora set up a boxing ring in its parlor and invited industry representatives to put on gloves and have their picture taken with the Champ.

The Dempsey-Firpo model sold well enough to justify going ahead with more. At the 1966 trade show Aurora had six sports figure diorama kits spotlighted in a black velvet case: Willie Mays, Dempsey-Firpo, Babe Ruth, Jimmy Brown, Johnny Unitas, and Jerry West (860-865). Unfortunately, these very well done kits, in the words of Frank Carver, "died miserably." It turned out that most sports fans just weren't model builders.

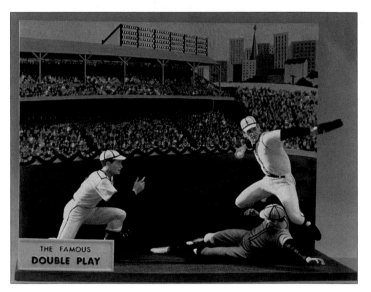

HMS sculpted a pattern for the Double Play, but poor sales of other kits in the Great Moments in Sports series scuttled its release.

The next line Silverstein planned came in response to an outpouring of letters from the public requesting a John F. Kennedy figure kit. The model of Kennedy (851) in his rocking chair with a portrait of PT-109 over the fireplace came out in 1965. It turned out to be only a fair seller, but Aurora announced seven other president kits in its 1965 catalog: Washington, Grant, Teddy Roosevelt, Jackson, Eisenhower, F. D. Roosevelt, and Lincoln. Only Washington (852) was actually produced. It was released in 1967 and, to quote Ray Haines, "sank right down to the bottom of the Delaware River!"

The Great Moments in Sports (863 $120-140) models built into great dioramas, but sports fans were not enthusiastic about them.

John F. Kennedy (851 $140-150) was
memorialized in plastic by Aurora.

Alfred E. Neuman (802 $70-80) had four different
poses and slogans. *Photo by Pat Mazza.*

Silverstein's next project was more offbeat. He went to
see William Gaines, publisher of *Mad* magazine, and received permission to make a kit of Alfred E. Neuman (802).
Ray Haines went to Madison Avenue to gather information from the staff, which he found to be "all wacky." In
October of 1965 the "What, Me Worry?" model hit the
stores. Perhaps he should have worried. Silverstein discovered that his kit was "a total bust . . . an absolute disaster."
The distributors bought it, but the kids didn't. Analyzing
the kit's failure, Silverstein observed that his own research
showed that figure kit builders were either younger than
13 or older than 16. *Mad*'s readers were 13 to 16. "I had
created a product for which there was no market."

Two other figure kits that Aurora brought out in 1965
can not be blamed on Silverstein. He was on vacation in
Florida when Aurora agreed to make the Nutty Nose Nipper and the Wacky Back Whacker (806, 807). "I came
back," he explained, "and I saw them, and I was furious."

What had happened? Marvin Glass, president of the
nation's leading toy design company had paid a visit to
West Hempstead with drawings for a line of humorous
torture devices that were supposed to tie in with the monster craze. A few years earlier, Glass had approached
Aurora with an idea for a robot with a clear plastic body
and lots of moving gears that could be seen running inside. Giammarino turned it down, saying it would be
impossible to get all those plastic gears to mesh properly.
Glass then turned to Ideal, which developed the idea into
Mr. Machine and made millions from it. This time Shikes
prevailed over Giammarino's objections, perhaps thinking that if Aurora continued to turn Glass away he might
not bring them some good idea in the future. However,
Aurora agreed to make only two of Glass's designs.

On December 5, 1965 the *New York
Times* carried its first comic strips—an
Aurora advertising supplement.

HMS carved the patterns from Glass's conceptual drawings, Ferriot did the molds, and the models went into production. Kolodkin pronounced them, "a pain in the neck to mold." Frank Carver said they were hard to assemble because they had wood, rubber, and metal parts. "They were pieces of junk." People who bought them found this out, and returned them to the stores in record numbers. Aurora agreed to give unhappy customers the kit of their choice as a replacement.

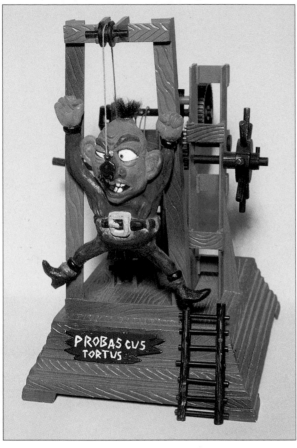

The Nutty Nose Nipper (806 $180-200) was an all-round disaster.

Another unusual set of kits, the Whoozis, had sales director John Cuomo as their sponsor. Cuomo was a great believer in slogans. As an advertising man, Cuomo felt that people could not resist stopping to read a catchy jingle. On his desk at the plant he kept a little ceramic figure holding various messages for his salesmen.

Susie Whoozis (201 $45-55).

He wanted a line of figure kits with each kit built around a slogan. Snuffy Whoozis (206), for example, reclines in a hammock over the proverb "Hard Work Never Killed Anyone, But I'm Not Taking Any Chances." It was hoped that people would buy the models because of the cute combination of a funny saying and a character figure. The Whoozis didn't catch on and lasted only a season.

The "Castle Creatures" were a pair of figure kits created in the same style as the Whoozis—a comic character combined with an offbeat saying. They were named for William Castle, producer of low-budget horror movies. The boys from Aurora had gotten to know Castle while in Hollywood during the Monster Customizing Contest and thought attaching his name to these models might boost sales. The Frog (451) was a comic creature sitting on a lily pad, while The Vampire (452) portrayed a cross-eyed female. They quickly joined the Whoozis in oblivion.

The figure kits that did catch on and rivaled the monsters as a smash success in the 1960s were the characters from comic books, television, and movies. These kits were a natural for Aurora since it had been advertising in DC comics for many months. These kits had the advantage of an enormous amount of free advertising in the form of the comic books, TV shows, and movies themselves. Comic books reached a peak of popularity in 1966, the high point in the "Silver Age" of comics. No kid had to be told who Superman or Captain America was.

The Castle Creatures (451, 452 $140-160) just weren't that humorous.

Superman (462 $140-160) led off a series of comic book heroes.

A Castle Creature that never reached the stores.

Silverstein went to Licensing Corporation of America and purchased the rights to make DC Comics' superheroes. Aurora's first kits, naturally, were DC's leading characters: Superman and Batman (462, 467). Although Aurora's Batman suffered from puffy facial features and poor part fit, he was a better seller than Superman. Abe Shikes son Stephan remembered that he liked Batman better because a boy with imagination could pretend he was behind Batman's mask.

1966 saw the first TV show figure kits: Illya Kuryakin and Napoleon Solo from NBC-TV's *The Man From U.N.C.L.E.* (411, 412) and Zorro (801), the only Aurora kit licensed from Walt Disney.

The Men from U.N.C.L.E. (411, 412 $200-220) were boxed separately, but the built models fit together into a diorama.

The original pattern of Batman (467 $270-300) had a base that was too small to keep him from falling over.

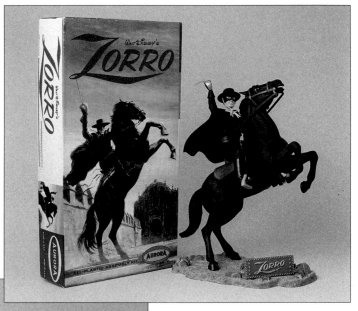

While most of Aurora's comic book heroes were from DC, three came from rival Marvel: The Hulk, Captain America, and Spiderman (421, 476, 477). One intrepid figure, Captain Action (480), was licensed from Ideal Toys.

Zorro (801 $250-300) was the only product Aurora licensed from Walt Disney.

Three of the comic book superheroes came from the pages of Marvel Comics (421, 477, 476).

Figure kits became so popular that Aurora went to the far reaches of the plant and dusted off some beryllium copper molds that hadn't been used since the 1950s. The US Marshal was revised and came back as Jesse James (408). Aramis got a new base and reappeared as D'Artagnan (410). The Gladiator pair also were put on new diorama bases and released as Spartacus (405) and the Gladiator (406).

Aurora's Batmobile and Batplane (486 $250-300, 487 $200-250) caught the wave of Batfever in the mid-1960s.

Spartacus and the Gladiator (405 $100-120, 406 $70-90) were 1965 reissues of the old 1950s gladiators, with new diorama bases.

A Batplane (487) was designed by HMS's artists based generally on the various batplanes that had appeared in the comic books. It arrived at the same time as the Batmobile. For the TV show's second season, Aurora added the Batcycle (810), the Batboat (811), and The Penguin (416). However, once the novelty of *Batman's* pop art graphics and spoofing of the comic genre wore off, the popularity of Batproducts plummeted. The TV show was canceled after three seasons. Aurora had a Batcopter, Batgirl, the Riddler, and the Joker in the works, but all these projects were abandoned.

Because Aurora already had a running start with comic book products, it was perfectly positioned to cash in on the Batman fad that swept the country following the airing of the *Batman* series on ABC-TV in 1966. Aurora rushed the Batmobile (486) through the production stages. Because the mold was cast beryllium copper, it could be made quickly and cheaply, and the kit was ready for shipment by the fall of 1966. Orders from distributors climbed to more than one million kits, so Aurora had a duplicate mold made and began running both sets of tooling twenty-four hours a day. Still they could not keep up with demand. It zoomed to become Aurora's best selling hero item, in Silverstein's words, "a giant runaway."

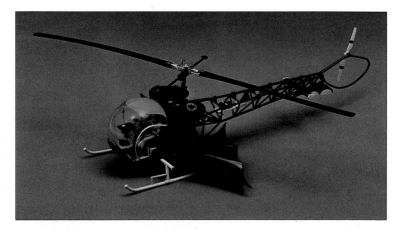

HMS worked-up a conceptual model for a Batcopter, but Aurora never produced it as a kit.

The career of the *Green Hornet*, a companion TV show on ABC, was even briefer—just the 1966 season. Today the program is remembered primarily for karate legend Bruce Lee's portrayal of sidekick Kato. Aurora's made a 1/32 scale model of Black Beauty (489), the Green Hornet's supercar.

Aurora's technicians were hired by ABC to build the pair of car models used by ABC in the show's special effects sequences. The Black Beauty and alter-ego Britt Reid's Chrysler convertible miraculously, but routinely flip places in the Green Hornet's garage. Derek Brand and newcomer Andrew P. Yanchus carved the 21" wooden special effects models in a hurry-up job that took just a month.

Wonder Woman (479 $380-400) is the rarest of the Superheroes because she was not re-released in the 1970s Comic Scenes series.

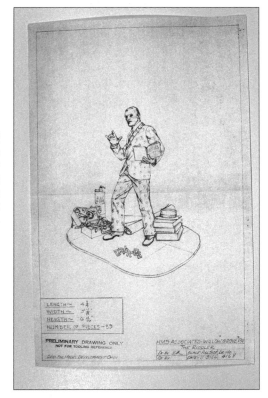

The design staff at HMS made drawings for The Riddler, but that's as far as development went.

The Seaview (707 $150-200, 253 $180-200) first appeared in the 1960s with Roy Grinnell box art; then in the 70s with art by John Amendola.

Andy Yanchus had begun working for Aurora in the fall of 1965. A Brooklyn native, he grew up with comic books and became an outstanding figure modeler. Yanchus, who monitored model quality for Ray Haines and designed slot cars for Derek Brand, recalled that one day Shikes stopped by his work bench and declared, "We should charge you to work here. You're having too much fun." Yanchus declared: "Working for Aurora was never boring. The day-to-day job of getting new kits into production was constantly interrupted by sales presentations, trade shows, public displays, TV commercials, photo sessions, and an endless stream of other off-beat projects and unexpected emergencies."

Television shows led Aurora into the science fiction field. As public interest in monsters faded, fascination with fantasy science increased. During this period Aurora created some science fiction kits that have become collector favorites today. The first kit in this line was the nuclear submarine *Seaview* (707), which each week managed to elude various underwater dangers on ABC-TV's *Voyage to the Bottom of the Sea*. Two years later the *Seaview* was joined by a model of its scout craft, the *Flying Sub* (817).

Lost in Space, which ran on television from 1965 through 1968, followed the adventures of the Robinson family, shipwrecked on a lost planet with stowaway Zachary Smith and a Robot. Aurora's 1967 catalog offered two versions of the show's Cyclops monster and the Robinsons (419, 420). The third model from this show was the Robot (418).

In 1968 the spaceship *Spindrift* took off from the United States headed for London, but ended up in ABC-TV's *The Land of the Giants* for two television seasons. This show inspired two kits: the Giant Snake (816) diorama and the Spaceship (830) model of the *Spindrift*.

Dr. Dolittle and the Pushme-Pullyou (814 $70-80) is a charming little model.

Four classic sci-fi models: the Cyclops from *Lost in Space* (419 $300-400), the Voyager from *Fantastic Voyage* (831 $400-450), the Spaceship from *Land of the Giants* (830 $160-200), and the UFO from *The Invaders* (813 $80-100).

Aurora found that kits based on television shows, movies, and comics were easy to make and usually sold well, in some cases very well. But not always. The kit of Doctor Dolittle and the Pushmi-Pullyu (814) took a while getting out of the pattern stage because actor Rex Harrison kept rejecting his figure model. "That's not me," he told Ray Haines. When the kit and Dolittle's ship *Flounder* (815) reached the stores, they did not sell. The movie turned out to be a flop, too. This resulted in the cancellation of other kits planned in the Doctor Dolittle series.

The Pushmi-Pullyu kit pioneered the use of snap-together construction that eliminates the need for cement and made assembly easier for youngsters. The model pieces had oversized locater pins that could be pressed snugly into plugs in their matching parts. This assembly feature would be employed in all Aurora's figure models introduced in the 1970s.

Aurora had ambitions to do a variety of Hanna-Barbera kits since their shows were big hits on television, but the end result was just one. Joe Giammarino and Ray Haines went to California to visit Joseph Barbera and got his permission to do some of his characters. The one to be completed was the Banana Buggy (832), an all-terrain vehicle driven by the Banana Splits bunch on TV. Toy push car and kit versions of the *Wacky Racers* and *Flintstones*, other Hanna-Barbera shows, didn't get past the test shot stage.

Among the most unusual Aurora kits are those from NBC's *Star Trek* TV show. AMT held the license to sell Trekkie models in the United States, and it created kits of the starship *Enterprise* and the alien Klingon Cruiser. In England, however, Aurora gained the rights to *Star Trek* kits. Aurora Great Britain imported and sold AMT's kits, in AMT boxes imprinted with Aurora logos (921, 923). Aurora Great Britain also sold a figure kit of Mr. Spock (922) shooting a phaser gun at a three headed snake. This figure kit had been carved by Bill Lemon in the United States.

The pattern for the Mad Barber (455 $1,000-1,200), an extremely rare kit issued only in Canada.

When licensing could not be worked out with the Johnny Weissmuller estate, Aurora dropped the Tarzan head and ape companion—and turned the model into Hercules and the Lion (481 $180-200).

The rarest of all Aurora figure kits are the Mad Professionals: Mad Barber, Mad Doctor, and Mad Dentist. Preliminary drawings for these kits were done by HMS in the summer of 1964, and the molds were subsequently made, but no kits were ever produced at West Hempstead. In 1966 Aurora Canada made a small run of the kits—so small that very few collectors have ever seen an original.

During the last years of the 1960s, Aurora contemplated returning to producing basic aircraft and ship models because figure kit sales were slipping and because Aurora wanted to reestablish itself as a mainstream model company. The R & D department started to pay more attention to research and detail in proposals for new kit subjects. A line of Russian military subjects went into development, including 1/72 scale models of the Yak-28 Firebar and MIG-25 Foxbat. Abe Shikes, who spoke fluent Russian, had paid a visit to the Soviet Union in 1966. Unfortunately, the new kit line would never see daylight because of changes in Aurora's management at the end of the 1960s.

The Mad Doctor (456 $1,000-1,200).

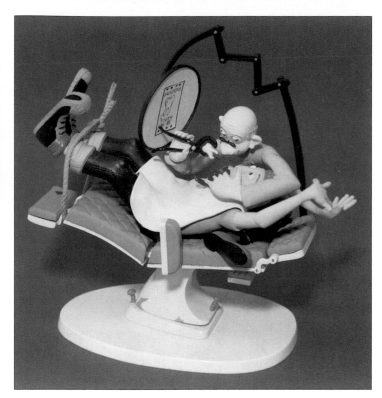

The Mad Dentist (457 $1,000-1,200).

The Aurora Empire

The 1960s had been a tremendous success for Aurora Plastics Corporation. The prosperity of the company could be seen in the expansion of its plant at West Hempstead. Aurora continued its practice from the 1950s of buying new molding machines to manufacture more models. Aurora never refused to make a model because it lacked the necessary equipment. It either bought another molding machine or farmed out the job to a contract shop. By 1970 Aurora had about forty machines in operation at Cherry Valley Road.

The factory expanded to accommodate them. A second floor was added to the front half of the main plant. Here, for the first time, major executives of the company got their own offices. Combined with other buildings in the neighborhood, this second floor gave Aurora 130,000 square feet of space. However, it was not enough, and in 1968 Aurora opened a three story building facing Hempstead Turnpike for Model Motoring. With the new building, Aurora had over 350,000 square feet of floor space. Aurora entered the decade with 235 employees, and ended it with 1,748.

In the mid-1960s Aurora moved to increase its market in foreign countries. Industry veteran Alvin Davis came in to head Aurora International, which opened an office in the Toy Center building in Manhattan. Davis's former company, National Model Distributors, had been one of the first wholesalers to carry Aurora kits, and Davis had originally put Aurora in touch with Mettoy to set up Aurora's distribution in Great Britain.

His first project at Aurora was to establish a subsidiary in Canada. Aurora products were already being exported to Canada, but advantages could be gained by having a company based there. Canada represented 10% of the North American market. By establishing a Canadian presence, Aurora could avoid import taxes as well as establish a base for exports to Great Britain and the Commonwealth nations.

Blackbeard and Captain Kidd (463 $230-250, 464 $50-70) are two of the most dynamically posed of Aurora's figure models.

In 1964 Aurora Canada began operations in a 10,000 square foot warehouse at 56 Brydon Drive, Rexdale, Ontario. Soon the company outgrew the original plant and headquarters moved around the corner to 31 Racine Road. The warehouse there started at 22,000 square feet and expanded to 80,000 by the end of the decade. The growth of the facility was a measure of the profitability of the company. Aurora sank just $25,000 into start-up costs for its Canadian operation and never had to invest another dollar there. The Canadian company had a great deal of independence in its operations. Al Davis recalled phoning Abe Shikes one time to get permission to produce a new product in Canada. Shikes replied: "If you have to ask me, what do I need you for?"

No manufacturing was done at the Rexdale plant. Kits imported from the United States in plastic bags were boxed by Rexdale's packaging machines. Those kits made in Canada were done by outside molding shops. Canada allowed tooling to be imported for a very low import duty, used for a while, and then shipped back to West Hempstead. Aurora Canada also printed its own instruction sheets. After a couple of early disasters, the Canadians learned that they could not do instruction sheets for kits heading to Europe in Canadian French. It had to be Old World French!

Selling in England had been difficult from the beginning because of peculiar English merchandising traditions. After Willem Thomas of Aurora Holland declined to supervise Aurora's interests in England, Davis went to England to end the contract with Mettoy/Playcraft and seek another distributor.

After failing in talks with Matchbox, he settled on Robert Chicken's company Model Hobby Products, located in Mebro Works, Cuckoo Hall, North London. Since Chicken exported Humbrol paints to the United States, Davis also set up a new corporation that made Aurora the American distributor of Humbrol paints. Chicken did not manufacture any Aurora products during the time he was their distributor in Great Britain, 1964-1968. Most of his imports came directly from the United States and some from Canada. In 1968 Aurora opened its own distribution company, Aurora Plastics, United Kingdom, at Green Dragon House, Croydon, South London. The plant remained there until 1971, when it was moved to Bexhill, a village on the Channel just west of the historic city of Hastings. In each case the plant consisted of just a warehouse with equipment for packaging imported products. Later on in the 1970s some kits were manufactured in England by contract shops.

All of its far flung activities in the hobby world caused Aurora's profits to soar during the 1960s. Although Aurora's financial success came primarily from sales of Model Motoring slot cars, plastic kits helped, too. In 1959 Aurora's total sales had stood at $5 million, but by 1969 they reached $30.7 million.

One reason for Aurora's growth was Abe Shikes' aggressive business attitude. Shikes pushed his executives hard, and he never let anyone forget that he was company boss, a role he played with gusto. A typical day at the plant for him might begin with a brief stint stacking boxes on the loading dock, followed by a walk to his office, during which he would "holler his way" through the plant. "That wakes them up, and it shows them I'm paying attention."

Shikes put an attendant in the ladies restroom to make sure the company's female workers didn't spend too much time powering their noses. He knew many employees by name and took the time to visit every department of the company. If a worker had his car repossessed, Shikes would make a phone call to the creditor and then send somebody (usually a big somebody) out to get the car back. At the end of the year everyone received a bonus paycheck. Although Shikes was naturally gregarious, he also believed that making your employees feel they were appreciated promoted good business.

In July, 1966 Aurora passed a milestone in its development when it joined the New York Stock Exchange. The listing of "AUR" on the Big Board marked Aurora's arrival as a major company. Shikes, Giammarino, and

Aurora developed a prototype for the Hardy Boys and their Rolls school bus, based on a cartoon TV show. The production tooling was never made.

Cuomo went down on the floor of the exchange with a small Model Motoring track layout to have their picture taken with Exchange president Keith Funston. At the time Aurora's stock was rising on speculation about a new product: Vac-U-Tron.

In the 1960s Americans woke up to the problem of air pollution. Aurora had plants in smoggy Los Angeles and New York, and in 1966 it had an invention that would combat air pollution from automobile exhaust.

Vac-U-Tron was developed at K & B. It was originally intended as an auto mileage booster, but it also operated to minimize emissions. The concept was to recycle and reburn exhaust emissions before allowing them to escape from the tailpipe. K & B tried Vac-U-Tron on UPS trucks in Los Angeles and gave twenty-five of them to the United States Auto Club. Both Shikes and Giammarino had them installed on their cars.

New York senator Robert Kennedy examines the Vac-U-Tron device installed on Abe Shikes' Cadillac.

"AUR" goes up on the big board of the New York Stock exchange. From the left: John Cuomo, exchange president Keith Funston, Abe Shikes, stock specialist Charles Schafer, and Joe Giammarino.

Success of the product hinged on a pending California law that would require all cars to pass emissions tests. Vac-U-Tron received certification as an acceptable way of meeting the law. Shikes boosted the product through a full page ad in the *New York Times* offering the device free to any New York department that would try it. Shikes even had his Congressman put in a plug for Vac-U-Tron into the *Congressional Record*. Unfortunately, California postponed its anti-pollution law, and that effectively killed Vac-U-Tron.

Despite the disappointment with Vac-U-Tron, Aurora was prospering mightily. In less than two decades it had moved from a garage in Brooklyn to the top of the world's hobby industry.

In the late 1960s Aurora began to tinker with the idea of expanding into the toy and game area. Shikes hired toy designer Anson Isaacson to come in and offer some fresh ideas. Isaacson had formerly been at Revell and Ideal. In one of Isaacson's first proposals, he suggested putting light bulbs into the movie monsters' heads to transform them into night lights. This concept was rejected as too expensive.

Derek Brand, who had launched Aurora into slot cars, also led Aurora into the game market. In his Model Motoring shop he hand made a pattern for an old English pub game called "skittles." The object is to knock little wooden pins over with a ball tethered to a rod inserted in the game's base. Giammarino and Silverstein modified the pattern and dubbed it "Skittle Bowl." During Christmas of 1968 Aurora sold $2 million worth of Skittle Bowl at $10 per item. In 1969 and 1970 it ruled stores as the best selling game in America.

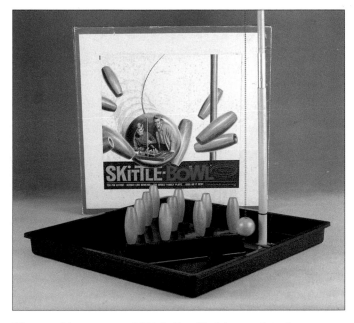

The smashing success of Skittle Bowl led Aurora into the games market—and caused the breakup of the old Aurora partnership.

The Old Guard Passes

During this time the company entered a two year period of crisis as dissension disrupted Aurora's executive offices. Part of the problem was the question of the company's future course: Shikes wanted to go into games and toys in a big way, and Giammarino did not. Shikes believed Skittle Bowl would lead the way for Aurora into the games market. Giammarino thought Skittle Bowl was a one-in-a-million lucky hit. A recession in the hobby industry that caused Aurora's profits to decline added to the tense atmosphere in the front office. An additional source of conflict was a clash of personalities among the company's founding partners. This would lead to the departure of all three of Aurora's founders.

The first to go was John Cuomo. At the end of 1967 he retired. He was 66 years old and at the age of retirement, but bad feelings between Cuomo and Shikes hastened his leaving. In 1971 Cuomo would pass away suddenly while traveling in Texas. Bill Silverstein echoed the sentiments of the whole hobby industry: "He was a sweet man. I liked John very, very much."

From the very beginning there had been friction between Shikes and Giammarino, but it had been the sort of abrasiveness that produced sparks of creativity. Now that friction consumed the old partnership. Giammarino was forced to leave Aurora in 1968. "I was fired from my own company," he declared. The board of directors, which had grown to include several "outside" men from Wall Street, voted for Giammarino's ouster, and Shikes went along.

Giammarino's departure shocked many of the company's old timers who regarded him as "Mr. Aurora." No product ever left the plant that did not first pass through his hands. Years later, when asked to explain Aurora's success, Giammarino mentioned hard work and said, "I don't know. I like to think it was because of me. My perseverance, stubbornness, and being a perfectionist. And picking the right products." He felt that getting involved with games marked the beginning of the end for Aurora. Early in 1969 Giammarino sold the bulk of his Aurora stock to a group of investors headed by New Yorker Charles M. Diker, who immediately replaced Shikes as company president. A new position "Chairman of the Board" was created for Shikes, but a year later in April 1970 he retired and followed Cuomo and Giammarino out of the Aurora picture. With his departure an era ended.

This store banner called attention to Aurora's Vietnam War-era models.

Chapter 7
Transition

Charles M. Diker became Aurora president with the intention of making it a modern company and leading it into new areas of growth. A cigar-smoking, thirty-four year old graduate of Harvard Business School, Diker held high ambitions for Aurora. He knew the Aurora name stood for quality products, but he felt that the company had never expanded its product line to its full potential. He also believed that Aurora's unsophisticated corporate structure held it back. Diker took on Aurora as a personal challenge to see what he could do with the company.

Charles Diker replaced Abe Shikes as Aurora president in 1969.

Diker had been a vice president at Revlon cosmetics and knew what a modern corporation was supposed to look like. Aurora had grown to be a large corporation, but in many respects it was still being managed like a garage company. Diker later commented, "Aurora had been run like a candy shop." He announced that he intended to correct the problem of understaffing, institute new organization and procedures, and increase spending on development of new products. He intended to dramatically increase sales volume. The only way to do this was to extend the product range from the limited hobby market into mass-market toys and games.

Essentially his analysis rang true. Aurora's product line needed an infusion of fresh thinking. Model Motoring's slot cars were miles ahead of the competition, but the system had not been upgraded for years. Aurora's plastic kit line had grown distinctly dated, with many of its stan-

dard models now more than a dozen years old. Moreover, the bloom had gone off the plastic model fad as young people's interests changed. If Aurora were going to grow, it would have to find new products.

The old time hobbyists at Aurora felt out of place under the new administration. As a result, almost all the second tier executives either left the company or were fired. At the top level, Aurora became virtually a new company. This transformation was symbolized by changing the name of the company to Aurora Products Corporation in May, 1970.

Former production manager Frank Carver expressed a common feeling among the departing old timers, many of whom came out of the 1940s hobby and craft tradition. He thought that the newcomers, with their MBA mentalities, didn't connect with the world of hobbyists and ten year olds. "In this business you have to have a heart for the product," he declared.

Carver also complained about the new management's emphasis on formal structure and procedures. He observed, "We had been very informal—unstructured totally. The new management said you can't run this company by the seat of your pants, but we did a hell of a job running it by the seat of our pants.... We made a lot of money. We were very happy."

The third area of criticism raised against Aurora's new leaders was excessive spending. Willem Thomas of Aurora Netherlands observed that the original partners always "did it simple," but under the new management "everything was expensive, expensive, expensive." Salaries were increased, new staff hired, new equipment purchased, and the formerly spartan executive offices were given new decor and, for the first time, carpeting on the floor. Speaking from the perspective of the molding room, Henry Kolodkin said, "We had an item that cost us a nickel to make. The way they wanted it made, it cost twenty cents."

The alleged excesses of the new Aurora were symbolized by a new sales showroom in the Toy Center building. It occupied the whole third floor and was the largest showroom of any company. The workers called it "Broadway" and observed, "When you flipped on the lights, the rest of New York City dimmed a little."

The new management might agree that the facts behind these criticisms were true, but insisted that change was necessary. The old leadership had already recognized that Aurora faced two choices: remain an old-time hobby company in a small market, or move up to the much larger toy and games market. Once the decision to follow the second course was made, a new toy-oriented staff, a much larger staff, and big investments in R & D, engineering, marketing, and sales had to be made. President Diker defended the expense, saying the results were worth the up-front costs. "You need to create excitement . . . make yourself the talk of the industry."

If the potential for gain was greater in the huge toy market than in hobbies, the danger of disaster also loomed larger. As Dick Schwarzchild explained: "Toys are a fashion industry. For every winner, you have to amortize a lot of losers. . . . Whether its Hollywood or cosmetics or the toy industry, you're going to have a hell of a lot of losers for every winner. The question is, how many losers can you afford?"

Aurora used celebrities like Don Adams to promote its games during the 1970s.

Diker appreciated the risks, but felt that once committed to its new course of action, Aurora had to press ahead aggressively with a major effort to establish a "critical mass" in the games and toy market that would command the attention of distributors, retailers, and the public. However, he did not want to abandon Aurora's secure place in the hobby world. The concept that he came up with to bridge the gap between hobbies and toys was "family." All Aurora products would have appeal to both sons and fathers—and could engage the whole family in fun activities.

In order to finance expansion Aurora needed an infusion of outside capital. Twice it attempted to sell issues of new stock on the open securities market, but failed to attract investors. The next strategy was to find a larger corporation that could furnish the needed financial muscle.

Ideal Toys and General Host, a craft company, gave Aurora some thought for a while, but on November 24, 1970, Nabisco announced it would purchase Aurora. The agreement hit a snag when Aurora reported a loss for 1970 totaling about $1.5 million. The purchase price was renegotiated down, and the buyout reached completion on May 28, 1971. Under the new ownership structure, Nabisco placed one executive in Aurora's offices to keep watch over its new acquisition, but otherwise left Aurora alone to manage its own affairs.

One sign of new thinking at Aurora was a move into the crafts field—the domain of women and a rapidly growing market. Craft shops were popping-up all across the country. Charles Diker met veteran craftswoman Aleene Eckstein at a hobby convention, and she agreed to sell her company to Aurora. The first new products to come from Aurora's partnership with Aleene were "Forged Foil" animals. Three of the Wildlife series kits—the Buffalo, White Stallion, and Cougar (445, 446, 447) were made into craft kits with the intention of selling them to women. Packaged in the boxes along with the plastic model pieces were a tube of Aurora plastic cement, Aleene's Tacky White Glue, Aleene's black antiquing paint, and several sheets of foil. The idea was to build the plastic kit, then cover it with foil to make it look like a cast metal figure.

As it turned out, Aleene had only a brief association with Aurora. In September, 1970 Eckstein left and formed another company of her own. (In 1977 she would reacquire the craft line when Aurora was sold, and she still appeared on television demonstrating crafts into the twenty-first century.)

If crafts turned out to be a dead end for Aurora, games did not. Within a few years Aurora's extensive lineup of tabletop games placed it second only to Milton-Bradley in the games market. Aurora's AFX slot cars were booming and even the toy department registered some hits. Under new management Aurora did become a much larger, more sophisticated company. Gross sales skyrocketed from $30 million in 1969 to $70 million in 1975. However, it was no longer a profitable company. The old Aurora never had a losing year; the new Aurora lost money every year.

The Forged Foil Cougar (447 $50-60) included the original Cougar model (without the fawn), a new rectangular base, two-sided gold-silver foil, and antiquing paint.

Chapter 8
Aurora Classics

The transition years in company management from 1968 to 1970 witnessed some hard thinking in the plastic model department. Models had taken a back seat to slot cars nearly a decade earlier, and then games pushed plastic kits even farther into the background.

Aurora's new Vice President for R & D and Product Development was Walter Moe, formerly of Ideal Toys. Moe realized that the growth curve for model kits had flattened out after years of growth. "By the 1970s hobby kits were a tradition, not a business, at Aurora," Moe asserted, "but we couldn't get out of it." However, he would not invest much more money in new kit development because the cost of new tooling was too great compared to the prospects for payback. The only advantage models had was that they sold steadily without much advertising. During the 1970s Aurora derived only about fifteen percent of its revenues from model kit sales.

Director of Sales Promotion Dick Schwarzchild summarized the thinking among Aurora's executives: "A very good case can be made, showing in black and white that in getting an adequate return on your investment, the hobby kit is an unprofitable investment. The cost today for a plastic mold is $35,000 on up, and getting that money out if you sell only 100,000 or 200,000 kits is impossible. When it cost $12,000 or $15,000 to make a mold, you could afford to put anything you wanted into the line. In our case, we could take a gamble with a Frankenstein or a Batman, both of which were very, very successful, or we could put out *Mad* Magazine's Alfred E. Neuman, which wasn't. But there wasn't anything to it then. If you hit, great! And if you didn't, the costs weren't such that you were going to hurt."

Schwarzchild showed an itemized list of model kit sales to Aurora's new men. Many kits were selling less than 5,000 units a year. Some were under 1,000. Schwarzchild explained that it was "simply unprofitable in terms of warehousing, in terms of packaging and handling, and in terms of even having it on your order sheet." The models offered in Aurora's catalog would be pared down from a high of 284 in 1967 to 100 in 1976. Figure kits took the biggest hit.

Aurora's toys division came up with some interesting products, such as a set of giant puzzles that could be framed and hung as posters.

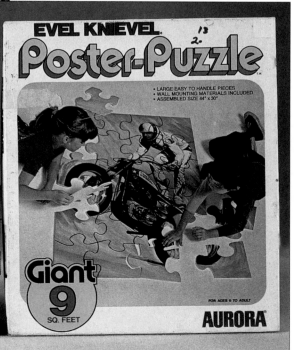

Having made his criticisms of plastic models, Schwarzchild backed off and said that models were a staple item in the hobby industry and would always be around. The problem was just finding the right market and matching it with a good product.

When he became president, Diker put all model projects on hold, except for those already at the tool makers. Only three new aircraft kits made it to the stores in 1970: the C-141 Starlifter (376), the A-7 Corsair (395), and the DC-10 airliner (366). The DC-10 suffered from lack of detail and poor part fit because the mold had not been made by Ace. Aurora was bringing out so few new kits that its relationship with Ace ended.

The artist for all three of the new aircraft kit boxes was Harry Schaare, an illustrator with long experience in commercial art. Born in Jamaica, New York, in 1922, Schaare grew up building wood and tissue model airplanes. Then during World War II he piloted the real thing, C-47 Dakotas in the European Theater. He had trained to be an architect at NYU, but following the war he studied art at Pratt Institute in Brooklyn and settled down to life as a commercial illustrator on Long Island. His bread-and-butter assignments were covers for pulp Western novels and men's magazines such as *True* and *Argosy*. Like John Steel, his favorite medium was casein—because it was water-based, easy to apply, and dried quickly.

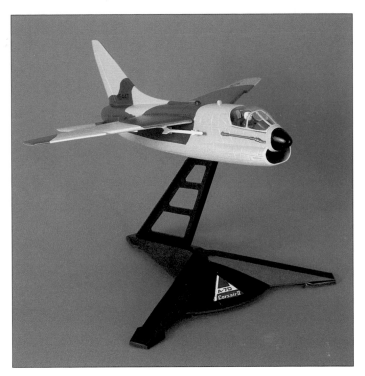

The A-7 Corsair (395 $25-35) showed that Aurora could produce an excellent quality model airplane.

Harry Schaare's new box art for the Japanese Sub I-19 (728) shows the looser, modern style Aurora aimed for in the 1970s.

Schaare became one Aurora's more productive, although less known, illustrators. He left most of his art unsigned, thinking that doing packaging illustration in the "toy" field would hurt his reputation in art circles. Since Aurora paid him only about $300 for each painting, they could not expect to get his most refined work. Schaare's box covers included many of the Vietnam era helicopters, the 1970s editions of the tanks, some of the "glow" movie monsters, and many figure kits. Probably his most striking creation is the Snake diorama from *Land of the Giants* (816). His last work in model box art was the *Planet of the Apes* series for Addar. In the mid-1970s he went West searching for better paying clients and became one of the nation's best-known Western genre painters.

What boy could resist Schaare's dramatic invitation to buy the *Land of the Giants* (816) snake scene?

In 1970 Aurora gave lots of publicity to its elaborate model of the Sea Lab III (721). This undersea station had been designed to test man's ability to live beneath the sea. It was a nice kit, with a side panel that lifted off to reveal the interior. Aurora even ran a "Man in the Sea Contest" to drum up interest. Entrants had to send in a box end along with their essay: "I'd like to live for two weeks on the ocean floor in Sealab III because..." Winners received water ski and scuba equipment.

Contests were designed to both sell model kits and attract customers into hobby shops.

Aurora's R & D people were already at work on other undersea projects like Jacques Cousteau's ship *Calypso* and his diving saucer, an Aluminaut sub with F-111 wreckage, and other experimental research subs. However, an agreement could not be worked out with Cousteau, and poor sales of the Sea Lab scuttled the rest of the series.

Aurora's final new ship model came out in 1971— the innovative Soviet helicopter carrier *Moscow* (722) that had just been in the news. This model marked the end of Aurora's relationship with Ray Haines and HMS—and to save money the mold was cut by a company in Canada. The *Moscow* made a fitting capstone to the HMS model tradition: simple, inexpensive, and the first model of a new, unusual subject.

Because of the new management's reluctance to investment in fresh model tooling, Aurora introduced only four new kit lines in the 1970s, and three of them were aimed more at the toy market than at serious modelers. The rest of its work in plastics consisted of simply revising and repackaging old products from the 1950s and 1960s. As Andy Yanchus, one of R & D's serious modelers declared: "Aurora was burdened with a succession of flashy gimmicks that gave the impression of great, new things from Aurora, but, in actuality, did nothing to improve the quality of the final product or the image of the company."

The first sign of product tinkering was conversion to the "square look" in kit boxes. In 1969 some kits were taken out of the rectangular, "flat" boxes used throughout the industry and put in larger, more cubical boxes. President Diker felt that the old box format wasted the top panel art when stacked on store shelves, while the square boxes presented a larger side panel to shoppers. The new boxes were white with clear graphics designed to stand out in stores. Diker also thought the company's oval logo had a "1950s look" and replaced it with a large capital "A" logo or a block letter AURORA.

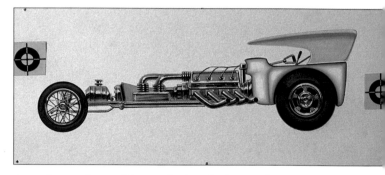

Aurora's new "square" boxes displayed a large side panel when stacked on shelves. This is the art for the 1971 reissue of Casper's Undertaker show dragster (570).

Aurora intended to issue several more kits in an underwater series if the Sealab III proved to be a success. However, the Aluminaut, with F-111 wreckage, and the Aquanaut never surfaced in hobby shops.

The large square boxes created the problem of finding something more to fill up the empty space inside the boxes. Critics of the toy and hobby industries had long complained of small items deceptively packaged in large boxes. The answer for some tanks and aircraft was a vacuum formed display base. However, the final answer was to largely abandon the square box idea. Aurora discovered that the big boxes added more in shipping costs than they gained in sales. Also, retailers disliked the larger boxes because they took up so much precious shelf space. In 1976 box size reverted to the smaller standard flat boxes.

The first kits to receive the square box and revised molds treatment were seven of the 1/32 hot rods, which appeared in 1970 as the "Scene Machines." The original hot rod models were so transformed by their hippy get-up they were almost unrecognizable. The old '32 Skid Doo (554), for example, became the Peppermint Fuzz (592) in pink plastic with a machine gun mounted on top and a "Hog Chopper" motorcycle resting on its running board. The decal sheets for these cars cost about eight cents, compared to one cent for most kits. Each kit came with a "cool" couple dressed in mod clothing. The young man in bell bottom pants with the Boob Tube (593) was inspired by Stephan Shikes, who worked briefly for his father's company on this project.

The Scene Machines led indirectly to the hiring of a new product manager: Jim Keeler. In the early 1960s, while serving an R & D man at Revell, he had led that company into development of ultra-detailed model car kits. Keeler happened to be shopping at his neighborhood supermarket in Utah, where he discovered a Peppermint Fuzz kit. He took it home, thought it looked silly, and telephoned president Diker to tell him Aurora could

do better. A few days later Keeler was flown to New York, hired, and given a chance to show what he could come up with. (Keeler unexpectedly found himself head of the slot car department, too, when Derek Brand, saying that the new managers held too many meetings, quit and went home to California.)

Arriving at West Hempstead, Keeler found a project called "Monster Scenes" already underway. The concept followed top management's idea that all company products should be slanted toward the toy market and have "play" potential. Individual kits in the set could be placed together with interlocking bases into a dungeon laboratory diorama. The "Scenes" concept also worked as a marketing device to separate Aurora's kits from those of all the other model companies.

However, it should have been evident to the R & D men that they were playing with fire, because some of the diorama kits were torture devices. They typified the sort of things that appeared in Warren magazines like *Creepy*, and, fittingly, one of the mini-monsters was Vampirella, a Warren character. However, the project's lead idea man, Andy Yanchus, tried to play down the horror aspects of the kits by turning the instruction sheets into comic books full of camp humor.

The marketing department, however, had different ideas and wanted to exploit the shock aspects of the kits, labeling the boxes with an "X" rating. At the HIAA convention, the Aurora parlor was hosted by women in low cut, slinky Vampirella gowns who drew attention to the aspects of sex and violence in the series. The reviewer for *Playthings* noted that "ghoulishness" from TV and movies had started creeping over into toys.

The old fire truck hot rod reappeared in hippie form in 1970 as The Wurst (596 $35-45).

This display was sent to hobby shops to show how the various Monster Scenes kits fit together into a whole play set. *Photo by Pat Mazza.*

Vampirella (638 $150-200) was the lead character in Monster Scenes.

The summer following release of the series the *New York Times* (July 12, 1971) ran a front page story, accompanied by a photograph of the Pendulum model, reporting that parent and feminist groups were concerned about "sadistic toys" which were "psychologically harmful." Nabisco's executives talked things over with the men at Aurora, and decided to take the Pendulum and Hanging Cage out of production. The controversy heated up again in November when a dozen women from NOW and Women's Strike for Peace began picketing Nabisco's headquarters on Park Avenue. Their signs read, "Sick toys for children make a sick society." At West Hempstead Dick Schwarzchild got a phone call from Nabisco: "We've got women with placards marching up and down in front of our building. What do we do?" Schwarzchild immediately replied: "Pray for rain."

Schwarzchild and Diker felt that kids viewed monster items very differently than adults and doubted that there would be protests had Aurora not been owned by Nabisco because pickets would have gone unnoticed on Cherry Valley Road. In fact, Schwarzchild observed that while Aurora had received some letters protesting the kits, it had gotten far more letters suggesting additional awful items for the Monster Scenes line.

"Monster Scenes" kits featured snap together 1/13 scale figures with movable arms and legs standing on tiny bases. These toy-like figures, with their stiff postures and lack of detail, just didn't measure up to Aurora's long-established figure kit standards. The four character kits were Frankenstein (633), Vampirella (638), Dr. Deadly (631), and The Victim (632), a scantily clad female, of course. The "grisly equipment kits" were Gruesome Goodies, the Pain Parlor, the Pendulum, and the Hanging Cage (634, 635, 636, 637).

The smaller, snap-together figure kits were not well received by the men who created them. Work on the patterns was divided between sculptors Ray Meyers and Bill Lemon, who had returned to doing work for Aurora. However, Lemon disapproved of the small, snap-assembly kits. "That changed everything. It made the figures mechanical. As far as I'm concerned it ruined the kits...I hated them. I know Ray hated them, too. It was so much easier to make them the other way. You just really couldn't do too much with them."

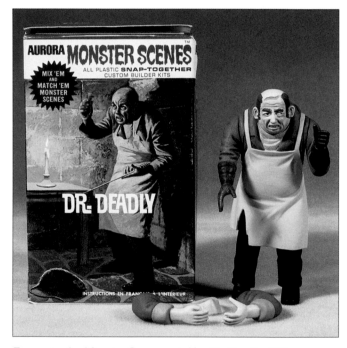

Figures in the Monster Scenes set, like Dr. Deadly (631 $100-110) came with alternate sets of arms and legs.

Nabisco, a conservative company that wanted to project a wholesome image, announced that the whole series would be discontinued. The kits remaining in inventory were shipped off to Canada. Included were three kits that were never released in the United States: Dracula, Dr. Jekyll and Mr. Hyde, and The Giant Insect (641, 642, 643). In Canada, The Victim received a new name, Dr. Deadly's Daughter.

The Victim (632 $100-110) was reissued in Canada as Dr. Deadly's Daughter.

Two other Monster Scenes accessories were in the pattern stage when the ax fell: a dungeon and an animal pit into which The Victim could be lowered. Still in the drafting stage were a hero, a second victim, an executioner, and more dungeon equipment. Keeler estimated that the tooling for the ill-fated series cost Aurora between $350,000 and $400,000.

The Giant Insect (643 $800-1,000) appeared only in Canada. The model pictured is a resin reproduction.

Eight kits were released in 1972, six in 1973, and three more in 1974. The best selling kit in the series was the Saber Tooth Tiger (733). Also joining it at the top of the sales chart were monsters embodying action and danger: the Flying Reptile, Allosaurus, Cave Bear, and Giant Bird (734, 736, 738, 739). The Allosaurus was the only one in the whole series not sculpted by either Lemon or Meyers. The Cave Bear had been developed earlier as the Grizzly Bear for the Wildlife series, but it did not go into production until Aurora revised it for Prehistoric Scenes. The human figures did not sell as well as the animals, and the background scene kits were the least popular.

Tyrannosaurus Rex (746) emerged as the final and the largest of the collection. It had been designed by Dave Cockrum, an artist at Marvel Comics, and sculpted by Lemon, who declared: "Gosh, he was so nice!" R & D head Walter Moe took an interest in this model. He had requested authenticity in the other kits, saying, "nature is spectacular enough." But in T-Rex he requested teeth twice the real size, thinking "kids relate to big, long teeth!"

The year after the Monster Scenes debacle, Aurora's model makers redeemed themselves with the best-selling kit line in the industry—"Prehistoric Scenes." The basic idea of this product was to make something easy to assemble, targeted at younger kids, and sold primarily through the toy departments of big chain stores such as K-Mart. Dinosaurs were just the ticket. Each came on a plastic base that interlocked with that of other beasts in the series and with "scenes" kits like the Cave, Tar Pit, and Jungle Swamp (732, 735, 740). They had moving parts so that they could be played with. Another feature intended to grab kids' attention was the selection of some "outrageous" colors of plastic, including red-orange. Mike Meyers said Aurora wanted to get away from the idea that dinosaurs were just "big gray beasts," and he observed that the choice of bright colors actually conformed to some recent scientific thinking on the subject.

Who cared that Cro-Magnon Man (730 $70-90), Neanderthal Man (729 $70-90), and Ankylosaurus (744 $130-150) didn't live on Earth at the same time?

Prehistoric Scenes gave kids some nifty ancient beasts with which to stage combats. The Triceratops and Allosaurus (741 $80-90, 736 $60-80).

By 1974 sales of the series were plummeting. The Saber Tooth Tiger went from 163,000 units in 1973 to 53,000 in 1974. That totaled a lot of kits by hobby shop standards, but fell short of K-Mart standards. Keeler felt that "the biggest mistake I made" was not marketing the Prehistoric Scenes as preassembled toys rather than model kits.

As usually happened in these cases, some models died on the production line. Dave Cockrum draw up plans for a large-size Stegosaurus that went to Bill Lemon for sculpting. "I always liked that one," Lemon recalled later. "The dinosaur with the big flaps all over it." A new, larger King Kong model would have been a natural match for Tyrannosaurus, but Kong remained only a creature of imagination. After 1976 Prehistoric Scenes went out of the catalog.

Aurora had an opportunity to bring out another figure kit series, but passed up the project. In the early 1970s, Anson Isaacson left Aurora to head Marvin Glass's toy design company in Chicago following Glass's death. Isaacson suggested to Aurora's managers that they do a line of figure kits based on the *Planet of the Apes* movies. The proposal went on to Keeler for his opinion, and he felt that figure kits seldom sold well: "Kids want to make a model of an airplane, not a model of the pilot." He intended to steer Aurora away from figure kits and other offbeat subjects into basic "hardware" kits: aircraft, ships, and cars. The higher-ups at West Hempstead agreed and turned down Isaacson's offer. (Shortly thereafter, while talking on the phone with Bill Silverstein, Isaacon would be shot and killed by a distraught Glass employee.)

Aurora by Another Name

In 1972 a group of the old-timers formerly associated with Aurora got together to form a new company and continue doing what they had been doing for twenty years at West Hempstead. The originator and president of the new company was ex-production manager Frank Carver, but the primary financier of the venture was Abe Shikes. It was no coincidence that Addar was another company whose name began with the letter "A." Shikes' association with the new company was not announced, but many people within the industry suspected it, and this helped Addar gain credibility with suppliers and distributors.

Addar's first products were the old Gowland ship-in-a-bottle kits that had been around since the early 1950s. The molds were leased and the kits produced at some of the contract molding shops that handled Aurora's overflow production. The packaging of the kits was done at Addar's small headquarters in Brooklyn. Distribution was handled by the same companies that Carver and Shikes had done business with over the years.

One morning Carver got a phone call from Shikes asking him to meet him for breakfast. Joining them was Anson Isaacson, formerly advisor on toys at Aurora.

Isaacson showed them photographs from a sequel to the 1968 movie, *Planet of the Apes*. He explained that he owned the rights to make plastic kits based on the characters in the movie, and Aurora had just turned him down. Would Addar be interested? Shikes said yes.

Cornelius (101 50-60) was one of the now-classic ape models issued by Addar.

Carver went to work on the production sequence just as it had been done at Aurora, using the same people who had served Aurora. Ray Haines supervised the mold making process. Bill Lemon carved the head parts for the figures, while sculptors at HMS finished the bodies and bases. Ferriot Brothers made the molds. Aurora artist Harry Schaare did the box art. Jim Cox drew the instruction sheets. George Burt printed the instruction sheets and boxes. The old team had returned to business. Addar was more Aurora than Aurora.

The Planet of the Apes kits were a great hit. Addar sold $2 million worth of them in a short time, keeping the machines running constantly to meet the demand. The series consisted of seven 1/11 scale figure kits and three "scenes" in the Gowland bottles.

The next set of kits done by Addar were also originals: three models of motorcycle dare-devil Evel Knievel created by HMS. However, the rest of Addar's growing line were from molds that Aurora had taken out of its catalog. Shikes went to West Hempstead and arranged to lease the tooling for eight of the old Comet aircraft kits. In 1977 Addar announced the forthcoming release of a number of Aurora's most famous kits, including early 1/48 scale aircraft like the Lockheed VTO and the F9F Panther, six 1/25 scale sports cars, and several of the sailing ships. Unfortunately, these kits never came out because both Aurora and Addar soon went out of business.

Addar lost Carver in 1974, and then Abe Shikes decided to retire a second time and move to California. With his departure Addar folded.

In 1987 Abe Shikes, a man who had never been sick a day in his life, a man who enjoyed singing along with the musicians in ethnic restaurants, died suddenly of cancer. Back in New York, Nat Polk organized a memorial service to honor his memory.

K & B Revival

The Collectors Series by Aurora's California subsidiary K & B was an approach to recycling some of the classic World War I aircraft that had been taken out of the Aurora catalog. Aurora opened a hobby and crafts division at K & B in 1972 under the direction of slot car specialist Jim Russell. This new department needed products to sell. As Russell explained it, "We had the tooling for these old biplanes lying around; so we thought we'd clean up the molds and see if the model kits would sell." Aurora of Canada and Great Britain also sold the kits, but with the Aurora trademark.

The items chosen for the Collectors Series were ten World War I and two between-the-wars aircraft kits that serious modelers had come to miss. Indeed, the hobby community warmly received the series. The old tooling had been refurbished, and some of the raised markings for decals removed. New colors of plastic, and revised, more authentic, decals were included. The results were still limited by the shortcomings of the original tooling and lack of time and manpower, but evidence of another ten years of research on World War I aircraft clearly showed in the models.

The kits came packaged in the square format boxes, with great new art work by John Amendola, Boeing Corporation's chief artist. Jim Keeler had met Amendola a few years earlier when Keeler worked briefly for Boeing as a wind tunnel model builder. Keeler wanted to make Aurora's box art more accurate and detailed, and he knew that Amendola was "an absolute stickler for detail." He also had the prime virtues of a commercial artist: he worked fast and willingly accepted a modest payment for his labors.

Amendola described his labors to World War I model historian Brad Hansen: "I indulge in a lot of research when I do an airplane, and often wondered if anyone could spot the many details or even cared for that aspect of a painting. The research on the K & B covers ate up a lot of time which was not covered by the price of the box art, but I felt it was necessary. I perhaps could have fudged a lot of it, but I thought I'd do the best I could... . I researched helmets, jackets, and goggles and tried to paint them as accurately as I could. Color info on many of the models was hard to come by."

A second series of six planes, including the highly desirable Spad, were announced for 1974, and Amendola painted covers for them, but they never went into production. The series was canceled after the original run because the only people buying the kits were hard-core older modelers. For the younger generation, World War I was just too remote to be of much interest. Russell wanted to go ahead and sell just to hobbyists, but top management wanted large volume sales.

The K & B Collector's Series brought back some of Aurora's World War I model planes with improved tooling and great John Amendola box art. The Breguet 14A (1141 $50-60)

During the 1970s Aurora's line changed more often by deletions than by additions. Some of the beryllium copper molds simply wore out, and it was not worth repairing or replacing them. The sailing ship molds, which had always been hard to work with, were losing their sharpness. R & D man Tom West observed that some of the hulls looked like they were "molded out of peanut butter." Anzio Beach and Rat Patrol sold very well at the big toy stores until the molds simply fell apart from too much pounding over the years.

Some of the earliest airplane molds continued to soldier on into the 1970s. The old B-26 had been built to be big, not good, but it continued to be amazingly popular and outsold most of the newer, more detailed kits. The P-40 and its cohorts, the ME-109, Focke Wulf, P-38, and Spitfire remained steady sellers. Their plastic colors had changed. In the early 1970s Andy Yanchus received the task of narrowing the wide range of colors used in the kits and games. One of his decisions was to get rid of the ME-109's "garish metallic red" plastic. In 1975, when the costs of plastic made the kits no longer profitable at $1, the old warbirds went out of the catalog.

Aurora's car models were dropped from the sales lists even earlier because the company's line of 1950s and 1960s automotive subjects had grown outdated, and constant exposure had exhausted their market. The last car models to go were five of the 1/25 sports cars. They appeared in 1971 with new decal sheets that allowed boys to decorate them like World War II fighter planes—"Battle Aces of the Road." The XKE became a "Spitfire" and the Porsche a "Messerschmitt." This fiddling typified Aurora's attempts to find an inexpensive way to perk up sales. However, the car kits were too complex for the youngsters who might have been attracted by the airplane decals gimmick. They lasted only two seasons, and when they left the catalog in 1973, Aurora had not a single car model left for sale. However, Aurora had a project in the works that promised to return it to the car market in a very big way.

Aurora had never gone into drag racing models as other companies had. Its sole entries into the dragster-funny car field had been the Hemi Under Glass and Mercury Exterminator (680, 670). However, dragsters were hot selling kits in the early 1970s, and Revell was making a big impact on the market with its 1/16 scale funny cars.

Keeler's first love was drag cars. He had won a national model customizing contest in 1968 with his *Dodge Fever* dragster, and Aurora had even paid him $5,000 for the right to make a model of his prize winning Dodge. Although this project had never materialized, his superiors now gave him the green light to create some really super kits with lots of options for creative modeling. Aurora decided to jump in with the most ambitious single automotive project ever attempted by any company. This represented a radical departure for a company known for its .69 cent hot rods.

Keeler justified the project by saying, "I wanted to prove to the modeling world that Aurora could make a quality kit." His original concept was to have two or three car kits that would be supplemented by parts kits and a garage diorama.

The company's top brass liked the idea, but insisted that the series become totally a "scenes" line to distinguish Aurora's kits from the dragster car kits already being sold by Revell and Monogram. Thus all the automotive components were divided into separate kits and sold separately. The car models were broken down into body, chassis, and engine kits. Added to these were kits for a garage, tools and equipment, and drivers and mechanics. These could be built, taken apart, and rebuilt in various combinations. Anything a mechanic could do with a real car, a hobbyist could do with the "Racing Scenes" kits.

The last gasp of the ill-fated Racing Scenes (851 $210-230, 852 $210-230) featured two of the best model car kits ever produced.

Once the project got underway, Keeler went to California and recruited Thomas West to help. West had just graduated from GM's automotive engineering school and was working as a technical illustrator for car magazines. Keeler asked West to design the patterns in 1/8 scale so that minute details could be put into the molds. West drew most of the engineering blueprints, but the bodies were designed by Richard Ratkiewich, the same staffer who designed most of Aurora's AFX slot car bodies.

To correct what the California R & D boys felt was "typical New York lack of automotive feel" among top executives, the model makers invited Charlie Wilson, a racer on the drag circuit, to bring his *Vicious Too* Camaro funny car to the parking lot at West Hempstead. Corporate heads gathered to see what it was all about. As West described it, "This group watched in horror as Wilson fired up the car and splashed a header load of Nitro all over some very expensive suits. Before they had a chance to get out of the way of the now rumbling fuel motor, Charlie had moved into the bleach and executed his first burnout. A Keystone Kops scene developed as a panicked herd of normally sophisticated executives struggled to escape the noise and smoke." A second ear-splitting burnout, followed by the arrival of the local police, completed the lesson in dragsterology.

When Racing Scenes previewed at the 1974 HIAA Trade Show, the industry did not know what to make of it. Here was Aurora, a company known for hitting the lower price points in the market, selling very expensive car kits. Many distributors refused to take a chance on a product so radically different from the usual industry products. Separately the kits were not very appealing; so to have something worth building and displaying, a modeler had to buy several expensive kits. Retailers were reluctant to devote so much shelf space to an untried product. Even the packaging presented a puzzle. The HIAA convention judges awarded it a gold star, but West thought the box art work was "styled more appropriately for denture cleaners."

Before the kits could be sold, however, they had to be produced, and every stage of the process encountered problems. The patterns were made by a shop in Philadelphia and by craftsmen in Detroit. The final prototypes—unusually large and executed in polished acetate—were things of mind-boggling beauty. They were sent to Windsor, Ontario where the molds were made by Binder, MPC's tool company.

When the molds finally went into the machines, the one piece parts for the Chrysler 392 and Donovan 417 engine blocks proved to be a huge headache. Because each block was molded in one piece, with four distinct sides, it was necessary to attach sliding side cams to the major mold halves. These broke loose on an "hourly" basis.

Sales were slim that Christmas season. Unfortunately, Keeler's attempt to prove that Aurora could produce a quality product ended up, in his words, as "a great big debacle." He blamed Racing Scenes' failure partly on the separate packaging concept, but more on the retail market's reluctance to carry the kits.

To try to salvage something from the project, the Vega body and the Pinto body were packaged with a chassis, an engine, and several figures as "Super Scale" car kits (851, 852). The decal designs for the cars were created by two artists who painted real dragsters: Nat Quick and Ken Youngblood. However, these kits sold for twice the cost of Revell's comparable 1/16 cars. West estimated that perhaps as few as 2,000 of each kit were made.

The Super Scale *Voodoo* Vega (851) was a model of a real funny car sponsored by a team that included plastic car builder West. The *Voodoo* presented no threat on the drag strip, but was an award winning show car. It carried the Aurora slot car AFX logo on its sides—as does the model.

The Heroes Return

The year of the funny car also saw the return of Aurora's characters from the comic books in a "Comic Scenes" collection. Andy Yanchus and other superhero fans in R & D had been looking for a chance to revive these figure kits. The sales department liked the comic book figures because they could be mass marketed as toys.

A decade had gone by since these kits had been developed, and some of them needed updating. Batman, Robin, and Superman (187, 193, 185) lost the inscribed logos on their chests. These were replaced with stick-ons updated to the current comic book style. The nameplates that had come with the original issue kits were deleted. Robin, whose original face had been flat and featureless, got a new hairstyle, new mask, and handsome new face.

Packaging for the series was the work of Andy Yanchus. He established contact with comic book artist Dave Cockrum and worked out an arrangement for the box art to be done by the same illustrators who did DC and Marvel comics. Cockrum and his fellow artists formed a company to produce the art for Aurora. The side panels of Aurora's kit boxes were exactly comic book size. They stacked well on store shelves, showing the terrific art work done by some of the industry's giants. Included with each kit was an instruction sheet in the form of an eight-page comic book, complete with a brief story and a middle section that could be opened to form a backdrop for the finished model.

The original long box Superboy (478 $180-200) came back with some mold changes and a new box in the Comic Scenes (186 $80-100).

The retail industry liked this series very well, and it sold into the market strongly in 1974. At first sales in the stores were brisk, and all seemed well, but then, in the words of West, "sales absolutely stiffed." Why? It seems that early sales were going to hard-core comic fanatics, who bought out the stores, stocking up on their favorite characters. After that, the typical kid was content to buy toy figures. Robin (193) and Captain America (192) got into the stores in the fall of 1974, as the final issues in the Comic Scenes.

Comic Scenes gave Aurora a chance to make some improvements to the Robin (488 $140-160, 193 $120-140) tooling.

For figure kit collectors, the demise of the series turned out to be a great tragedy because Aurora had two brand new comic book heroes in the works that would have been among the best ever. Comic book artist Dick Giordano designed figures of Flash Gordon in a sword fight with Emperor Ming. The arms of the figures could be moved to a number of realistic positions. (Meyers sculpted Flash, Lemon did Ming.) The second model was Dave Cockrum's conception of The Phantom, straining to hold the collar of his wolf dog Devil who is lunging forward with just one rear paw on the ground. Lemon sculpted the pattern, which still exists.

Although the Comic Scenes were in the catalog for only two years, a fairly large run of the kits were made, ranging from a low of about 175,000 for the less popular kits to about 275,000 for Batman. Shortly after shepherding the Comic Scenes series through to completion, Andy Yanchus left Aurora to become head colorist at Marvel Comics.

In 1975 Aurora returned to the past in another way by renewing its connection with *Parents' Magazine*. Beginning in October, 1975 *Parents'* ran an ad offering a kit a month to anyone who joined the Young Model Builders Club. Surprisingly, sales through the Club were not at all bad, and it offered the model fans at Aurora a chance to reissue some of their favorite kits. Tom West got "a touch of satisfaction" from redoing the instruction sheet for the XKE sports car. Not too many years earlier, as a youngster, he had built this model and ended up with two leftover parts. Now he learned where those two parts fit and did the instructions to make sure everyone else would too.

This resin model of The Phantom was sculpted in the 1990s by Rick Wyatt of Action Hobbies. It is based on Dave Cockrum's design drawings for Aurora and comes close to matching Bill Lemon's original model pattern.

In all, perhaps as many as thirty kits were brought out through the Club. Kits in this series are neither very common nor collectible, but they are easy to spot because they were packaged in corrugated shipping boxes with the art work glued to the side. Revell had supplied kits to the YMBC before Aurora, and when Aurora's kit line was sold to Monogram in 1977, that company took over the

Parents mail order club. Thus today Revell, Aurora, and Monogram kits can be found in Young Model Builders Club mailer boxes.

Another classic set of kits, the Knights in Armor, were reissued in a couple of interesting ways. After 1971 the knights were deleted from the Aurora catalog, and the molds were shipped to Aurora Great Britain because the knights were popular in Europe. One man who regretted seeing the knights leave was Nat Polk. He contacted Aurora Great Britain and asked for a run of 15,000 each on the first four knights (471-474). These kits were shipped to the US in unsealed boxes without the little helmet feathers because including them would have raised import taxes to prohibitive levels! Knight issues from the Polks' run are distinguished by their typical European thin cardboard boxes and by a paste-on sticker reading "Made in England."

The second reissue of the knights also came from Great Britain. The R & D department at West Hempstead had planned to renew the knights by giving them new bases, electroplating them silver, and including antiquing paint with each kit. The executives in New York shelved the idea, but Aurora Great Britain picked it up. The new bases and antiquing paint were dropped, but the knights came out in England as the Knights in Shining Armour (881-886).

Searching for another set of kits that might be worthy of reissue, Keeler, who's first love remained car models, began work on reviving the Old Timers series of vintage automobiles. Shortly thereafter Keeler left Aurora, and Tom West assumed direction of the project. He went to the Long Island home of antique car collector Henry Austin Clark, Jr. to inspect and photograph the cars in his museum. The old molds were updated to give them more authenticity and to make them easier to assemble.

The original 1961 issue Stanley Steamer (573) had retained the bent wire hangers used in the old Hudson wooden kits to attach the fenders and running boards to the chassis. These wire parts were very difficult to cement into the correct position. The reissue of this kit replaced them with plastic hangers that locked into the correct position.

Textures for the canvas roofs and leather seats were etched into the molds, and details were inscribed on the chassis. Two of the cars were given authentic white rubber tires. The result was a high quality product. Tom West was a bit put off by those hobby magazines that simply dismissed them as "reissues."

In the early 1970s the normal flow of business at Aurora was interrupted by the world oil crisis of 1973-1974. Manufacturers saw the price of plastic skyrocket. Fears that some plastic and petroleum products might become completely unavailable proved unfounded, but such fears led to hoarding. President Diker later recalled, "We had to beg, borrow, and steal to get plastic. We had to cajole

the vendors." However, by 1975 prices had stabilized at higher levels and business returned to more normal conditions.

Having weathered the storm outside in the American business world, Aurora received another shock from the inside. Nabisco president Lee S. Bickmore, the man who had spearheaded the purchase of Aurora, was obliged to retire due to poor health. He had been an enthusiastic supporter of investing in Aurora's growth, but the men who followed him were more interested in bottom line profits. They deplored the annual losses at Aurora and decided to cut back on support for the company. This turn of events dismayed Diker. Unwilling to preside over a conservative retrenchment, he departed in 1975.

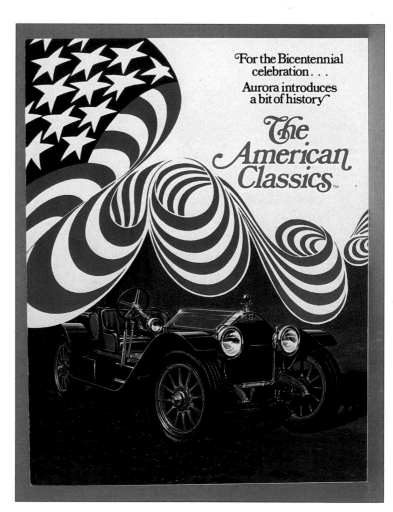

The Old Timers cars came back as American Classics in the 1970s.

Nabisco brought in a new man to lead Aurora in a new direction—or, actually, an old direction. The incoming president, Boyd W. Browne, declared that his job was, "quite simply to make the company profitable... In a word we're concentrating on what we do best." This meant cutting back on games and strengthening Model Motoring, but it included a small renaissance in plastic kits.

Browne came to Aurora from Mattel of Canada, where he had been president. He assumed his new duties on October 1, 1975. Forty-four years old, with a background in sales and marketing, Browne represented Aurora's last chance to survive.

Tom West remembered that conditions in the injection molding plant had deteriorated. The shop had become dirty, molds were being abused and even lost, kits were incorrectly packaged, and the second shift was "totally out of control." Worker morale and productivity had fallen decidedly lower than it had been in the 1950s. Front office men avoided setting foot in the place.

Andy Yanchus, a very serious modeler, later wrote: "All of the problems of the kit line had a common root. There just simply weren't enough people at Aurora interested enough in the hobby to ensure that the final product was an excellent one. Oh, there were some who cared. I was one of them. In the small Research and Development staff that I was a part of, were some of the biggest modeling fanatics you could find anywhere. It seemed we were at odds with just about every other department. We fought like hell to get better kits, and every little victory we had seemed like a major triumph. However, we never could win a decisive battle."

Improvements in the mold over the years made the Panther Tank (071 $15-20) a pretty sophisticated model by the 1970s.

Despite complaints from its model enthusiasts, Aurora produced some good kits in 1976 and 1977. Brown and West set up a five year plan to revive the model line by getting rid of "junk" and upgrading the remaining kits with the best sales potential.

The tank line, which had been high quality from the start and had received new packaging in the 1970s, came back in 1976 with another new look. Minor changes were made in details of the molds, and the tanks' armor re-

ceived just a little bit of texture that would, in West's words, "take away the shine." The four poorest selling kits were not revived: the 155mm Long Tom, 8" Howitzer, Churchill tank, and Swedish S tank.

Ten of the 1/32 scale Street Rods reappeared in bright new colors and configurations. For this final edition, the cavities to the mold sections that had produced the Scene Machines' figures and other odd accessories were shut off or filled in. The new issues received mag wheels and fat "Indy Profile" tires, plus some accessories such as roll bars. New pinstriping decals were made, and the first run of the kits included an "iron on" decal for the kit builder's t-shirt. One bad thing was that these kits did not sell for .49 cents anymore.

Many of the World War I aircraft which appeared in the 1977 catalog had been through the K & B Collectors Series reissue and now returned with more revisions and improvements. The first six Great War kits that had begun the series back in 1956 were still going strong. They received a surface texture reproducing the look of fabric, and all the raised decal locater markings were finally erased. All the kits were given two sets of decals representing real aircraft—a further benefit of more research into World War I aviation.

To make building these models easier, the separate interplane struts were molded in pairs that fit into slots into the lower wing halves. Project director West was relieved when the first test shots came out of the machines because he designed the strut revisions from hand sketches, not engineering drawings. They fit perfectly. "You certainly pulled this off," he thought to himself.

The old Albatross D-3 (104 $15-40) from 1956 returned as a D-5 (752 $12-15) in 1976 with a more rounded tail fin and one of the radiators removed from the top of the wing.

Unfortunately, the ground base and mechanic figures were deleted from the kits because their places in the molds were taken over by the parts that had formerly been molded in black in auxiliary molds. These kits also were plagued with sloppy packaging that resulted in missing parts.

The Science Fiction series of 1976 provided the most enjoyment for the R & D boys and is a favorite of today's kit collectors. Putting it together showed how something could be scavenged on a very low budget.

Included in this collection was the old *Impetus* Nuclear Airliner (129) from 1960, which might be considered Aurora's first science fiction kit, although at the time it was seen as a futuristic concept aircraft. In its new guise the Nuclear Airliner became the *Ragnarok* Orbital Interceptor (251), an idea based on a design by Yanchus. He conceived of a doomsday squadron battling invaders from outer space—thus the name *Ragnarok*, the final battle of the gods in Norse mythology. As released, the kit's fictional story line was brightened up a bit and *Ragnarok* was given an erroneous definition as "the Norse god of war."

Molding the kit's stand in black plastic, rather than clear plastic, was another of Yanchus's ideas, but the mold was still the old tooling. On the underside of the stand the company's former name, Aurora Plastics, remained.

Other kits received similar revisions. The Flying Sub (254) had just the color of its interior parts changed from silver to metallic blue. The dome of the *Spindrift* space ship (255) was changed to transparent red to match the color in the TV show. The *Seaview* (253) switched colors from black to gray, received some inscribed lines on its hull, and was mounted on the larger base formerly used by the unsuccessful Sealab III model. The Flying Saucer (256) from ABC-TV's *The Invaders*, a 1967 television show, had its underside domes switched from metallic silver to transparent red. It was placed on the ground base of the Dick Tracy *Space Coupe* (819), and three of the Tracy figures joined the original Invader figures.

Aurora introduced its final line of original kits in 1975, and, appropriately, they were remakes of one of its best selling collections, the Universal movie monsters. To keep down the static that had greeted its earlier monster kits, Aurora presented these as models of actors playing classic movie characters.

Monsters of the Movies turned out to be Andy Yanchus' last project at Aurora. He hired Marvel's Dave Cockrum to draw the three-view renditions of the figures that were used by pattern makers Ray Meyers and Bill Lemon to carve the model prototypes. For a while the R & D team worked along the lines of the "scenes" concept, developing diorama features like an old-time movie camera and cameraman, but since these elements would have added to the cost of the kits, ultimately it was decided to go with small bases for the figures. The new monsters were scaled in smaller 1/12 scale, rather than the original 1/8 scale.

Three of the kits in the new series were salvaged from the ill-fated Monster Scenes series of 1971: Dr. Jekyll, Mr. Hyde, and Dracula (654, 655, 656). They were joined by Frankenstein, the Wolf Man (who looked more like Lon Chaney, Jr. this time), and an underwater scene Creature from the Black Lagoon (651, 652, 653). The final two kits in the series were movie monsters from Japan's Toho Corporation: the flying monster Rodan (657) and the three-headed dragon Ghidorah (658).

The Monsters of the Movies did not become not major sellers, partly because public interest had shifted away from the old-time monsters to science fiction and partly because Aurora's distribution system to retailers had shrunken drastically.

The Monsters of the Movies Creature (653 $250-300) shows Bill Lemon's talents at their best.

The Dr. Jekyll and Mr. Hyde (654 $90-110, 655 $90-110) models were designed to be mirror images of each other.

Once again, some great figure model kits intended for the series died in the production pipeline. Dave Cockrum's Metaluna Mutant, Phantom of the Opera with his lady love, and Godzilla all had patterns made for them. So did a 1/12 scale version of Fay Wray—a figure intended to go with a new, larger King Kong. Cockrum also drew conceptual drawings of The Fly smashing-up his laboratory and The Mummy lurching past a statue of Anubis, but these seem not to have made it off the drawing board.

By 1976 things reached a crisis stage for Aurora's plastic kit department. The R & D men told management that unless they were willing to invest significantly more money in new models, Aurora should get out of kits altogether. The reply seemed to be, "OK, we'll drop plastic models." However, in the fall, word came down that Aurora would have a kit line in 1977 after all. By this late date it was impossible to do anything but continue production of those kits that had already been refurbished.

There remained, nevertheless, another way Aurora could expand its model kit line without incurring tooling costs, and that was by making a partnership with another company to utilize their products. Aurora had already done this in a small way.

The Metaluna Mutant from the movie *This Island Earth* never made it to stores, but the model pattern survived. In 2001 hobbyists Phil Ceperano and Al Reboiro used Ray Meyers' original pattern to recast just two dozen models in resin.

In 1975 Aurora sold a nifty set of tiny snap-together models imported from Australia, the Snap-A-Roos. These had been created as cereal premiums by the Australian firm, Rosenhain and Lipman. Aurora packaged them four to a box with the hope that kids would buy several of the low priced sets. Aurora had added a collection of kits to its offerings without expending a penny for new tooling. The Snap-a-Roos were sold in stores as toys, and only attracted modest attention.

Heller of France approached Aurora with an offer to make Aurora the American distributor of Heller's kits, and also those of the Italian company ESCI. The Polk brothers were Heller's current distributors, and Nat Polk warned Aurora's Boyd Browne that Heller's kits were too expensive and complex for Aurora's traditional youthful customers. Polk advised Browne to simply bring back some of Aurora's better old kits. Nevertheless, Aurora went through with the agreement.

This figure of the Phantom of the Opera's love interest failed to go into production. *Photo by Tom West.*

Heller and ESCI kits were imported from Europe in plastic bags and then boxed in the US. Tom West declared: "Heller was an appropriately named company because their complicated kits were hell to build." He and the artists of the model division revised and translated the French and Italian instruction sheets, despairing over the translations of some of the obscure French nautical terms used for the ship kits.

The Aurora/Heller and Aurora/ESCI "Prestige Series" ran to a total of 76 kits. The most impressive model in the collection was the massive and elaborate *Soleil Royal* (6550), a seventeenth century French ship of the line. Ten other historic sailing ships for which Heller was justly famous were included in the series. Heller also furnished a dozen World War II aircraft in 1/72 scale. The final Heller kit was a showpiece 1/35 scale Super Frelon helicopter that epitomized the most recent refinements in the plastic manufacturers' craft.

ESCI contributed six 1/9 scale motorcycles. The other ESCI kits were an extraordinarily vast collection of Second World War tanks, trucks, and cannon in 1/72 scale, accompanied by 13 boxed sets of fighting men, ranging from English paratroopers, to European partisans, to Russian Red Guards.

This venture into importing kits from foreign manufacturers was Aurora's last enterprise in the plastic model field. Months earlier, back in the summer of 1976, Nabisco had decided to sell Aurora.

How had Aurora—a leading company in plastic kits and still the world leader in slot cars—arrived at this sad terminus? Partly its problems had been generated internally, but the demise of Aurora also reflected difficulties besetting the entire hobby industry. In the 1970s there were fewer plastic model manufacturers, fewer distributors, and fewer hobby shops. With the Baby Boom over, there were fewer kids in the population. At Christmas in 1976 stores began offering a new toy for kids: games that could be hooked up to home televisions and played on the video screen. Youngsters were finding new ways to occupy their time.

It was a great loss to hobbyists that this new Godzilla did not join the Monsters of the Movies. The arms move to tear Tokyo Tower apart, and when the jaws open, Godzilla sticks out his tongue. *Photo by Tom West.*

The old hobby industry fraternity of the 1950s and 1960s disappeared. The HIAA became a trade organization dealing strictly in the crafts line. Cuomo, Giammarino and Shikes were only three among a number of hobby industry leaders who departed from the scene. Revell's Lew Glaser died in 1972. Monogram's Jack Besser retired in 1975, saying that the business just wasn't the same anymore.

For Nabisco, Aurora had been a disappointment. The new management had succeeded in increasing Aurora's gross sales to a peak of about $70 million in 1975. This was a little more than one-fifth the revenues of toy industry giant Mattel and a little less than half those of games market leader Milton-Bradley. Following the cutbacks in games, sales fell to $53.4 million in 1976. However, during these years Aurora's management had sacrificed profits to build sales volume. Nabisco's annual report for 1977 announced a total loss over six years of $25.9 million. Ultimately, it would end up being worse than this.

Nabisco decided that the most convenient way to divest itself of Aurora was to break the company down into its parts—a decision which doomed Aurora's chances for survival in its pre-1971 state. The plastic kit division was among the first to go. The sale of Aurora's plastic model assets to Monogram Models was completed in May of 1977, but the molding machines continued to run at West Hempstead until July to fill orders.

Nabisco sold Aurora's K & B division to Leisure Dynamics, but that company ran into trouble a few years later, and John Brodbeck repurchased his old company and continued to run it down to the 1990s. Today K & B occupies new headquarters in Lake Havasu, Arizona. They are the United States' leading manufacturer of model airplane and boat engines.

The last parts of Aurora, the games and AFX slot car divisions, were sold in November of 1978 to Dunbee-Combex-Marx, an international toy conglomerate that had recently purchased the English model company Frog and the American toy company Louis Marx. This new company ran into financial difficulties and declared bankruptcy in February, 1980. Nabisco, one of Dunbee-Combex-Marx's largest creditors, received only a small portion of the debt still owed it.

The Aurora AFX slot car trademark, which kept the name Aurora alive for many years, passed through several hands over the years and was picked up by Tomy corporation of Japan, although Tomy finally dropped use of the Aurora name in 1995.

Chapter 9
The Legend Continues

In 1977 Monogram Models of Morton Grove, Illinois, was the nation's second-largest model company, trailing only Revell. The executives at Monogram decided to acquire Aurora's large inventory of tooling in a strategic defensive move to take the molds out of circulation. With Aurora gone and its molds denied to any other company, Monogram would have one less company competing for shelf space in the stores. Monogram's leaders felt their company would benefit even if it didn't reuse most of the molds.

Later that winter, the molds, as well as reference material, art work, prototypes, and the kit archive, were cleared out, loaded into trailers, and placed on railroad flatcars. Passing through upstate New York on the way to Monogram's headquarters in Morton Grove, Illinois, the train derailed, toppling trailers and scattering molds across a frozen field. Once at the Monogram plant the molds were stacked in a warehouse where caked-on ice and mud were cleaned off. Five of the molds were damaged beyond repair: the Aero Jet Commander (85), Halberstadt CL II (136), Breguet 14 (141), Albatross C-3 (142), and Cessna Skymaster (279).

The molds that survived the trip were inventoried and those that were not worth repairing or did not fit the Monogram line were scrapped. Aurora had already destroyed many unused molds, such as the Guys and Gals, a few years earlier. Beryllium copper molds were particularly vulnerable to being broken up because of that metal's high scrap value. Monogram executive Robert Johnson later recalled there was a "feeding frenzy" of mold destruction, with minimum consideration of what was being lost.

Some of the molds were obsolete antiques that had only sentimental value, but many of the molds that were sold for scrap were modern and quite good. Some of the Racing Scenes tooling had been damaged, so all the molds in that series went to the junkyard. The *Seaview*, a favorite of today's collectors, was one of the first molds wrecked. Some molds simply couldn't be found in the warehouse when Monogram's product executives later went looking for them.

Monogram immediately put a few of Aurora's kits, including four of the airliners, into its line right away, and some ship and car models also became Monogram standards. The Prehistoric Scenes dinosaurs returned to the Monogram (and later Revell) catalog several times.

Model companies are reluctant to talk about their tool bank because it determines their future kit issues. Monogram has not released a list of the Aurora molds that still exist, but it is probable that only a few more than seventy Aurora molds have survived.

Revell-Monogram confirms that some molds exist for models that have never been reissued. Included in this group are all five of the original knights, Tarzan, and Captain America. The four great sailing ship molds also were saved from the wreckers. Ten of the 1/48 scale tank kit molds were kept, but only two, the German Panther tank and the Sherman have been used. From the hundreds of molds produced by Aurora over a quarter century, evidently only this remnant (and perhaps a surprise or two) survive.

Aurora's monsters and comic book heroes did not receive much attention from Monogram. Godzilla and Superman appeared briefly together as an strange duo in 1978. Then in 1983 the first four original movie monsters—Frankenstein, Dracula, Wolf Man, and the Mummy—returned with glow parts in modern square Monogram packaging. Then nothing.

The I-19 submarine (3103 $15-20) went into the Monogram catalog right away when Monogram acquired Aurora's molds.

Baltimore hobbyist Andrew Eisenberg got tired of waiting for Monogram to bring back the favorite models from his boyhood. So he formed the company CineModels and in 1991 paid Monogram to make a run of 5000 Draculas in the same packaging that had been used in 1983. Shortly after Eisenberg received his shipment of kits, Monogram released the first four monsters again—

this time in translucent "luminator" plastic that glowed in the dark. The next year two long-gone favorites, King Kong and the Phantom of the Opera, joined the luminator series. Aurora-starved hobbyists were delighted.

Although Monogram's reissues slowed sales of CineModels' Dracula kit, Eisenberg decided to take the progression of monster re-releases a step farther. He asked Monogram to make runs of the Forgotten Prisoner and the Phantom of the Opera and ship them to him in plastic bags. Then Eisenberg hired a printer to make reproductions of the original long-box versions of the kits. He contacted Tomy corporation about use of the Aurora oval logo, and they posed no objections. Eisenberg then investigated the legal status of the Aurora trademark and discovered that it had expired. So he filed an application to acquire the vintage Aurora emblem, and in 1999 the US government awarded CineModels the trademark.

Polar Lights

Meanwhile in the heartland of America, Tom Lowe, an enterprising businessman who had grown up with Johnny Lightning die-cast cars and Aurora monsters, discovered a groundswell of interest among men in the thirty-plus age range for the toys they had played with as youngsters. Aging baby boomers were diligently searching flea markets and garage sales for neat old stuff; prices for vintage toys on the collectibles market were soaring. Lowe decided there must be a better way to recover lost treasures, and he was pretty sure no established company would take a chance trying to revive these long-abandoned products from the past. In 1994 he founded Playing Mantis in South Bend, Indiana, to bring copies of 1960s-era Johnny Lightning cars to Wal-Mart and Toys "R" Us.

When Polar Lights copied Aurora's Bride of Frankenstein (5005 $18-20), it improved the original by adding clear plastic parts for the laboratory test tubes.

Years earlier, as a youngster living in Michigan, Lowe loved watching the *Munsters* and *Addams Family* shows on television. The spooky old houses featured on the TV shows aroused great fascination in his eight-year-old imagination, and when he discovered Aurora's Haunted House kit, he just had to build it—even if it meant asking for assistance from his older sister. Thirty years later, by the 1990s mint-in-the-box kits of the Haunted House were selling for more than $500. Lowe decided to find out if his favorite model could be brought back to life.

Lowe had already established a good working relationship with the agent in Hong Kong handling production of his Johnny Lightning cars. He sent him a generic model kit purchased at a hobby shop and asked if his supplier in China would be able to copy the parts from such a kit. They replied that they could, and gave him a figure on the cost. Lowe then bought an original Aurora Haunted House from a collector and shipped it to China. The kit parts served as patterns to machine-cut new molds. Since the original Aurora mold was cast in beryllium copper—rather than carved into steel—the new mold did not exactly duplicate the original tooling.

The economic calculus of bringing back an Aurora kit hinged on low-cost production in mainland China. Although Chinese mold fabrication techniques were far from state of the art, they could create a very high quality tool and then use that mold to produce kits at about one-fifth of what it would cost to manufacture the same product in the United States. Since sculpting of the model parts had already been done and paid for by Aurora years ago, the up-front costs of paying a sculptor and parts designer were not an expense. Thus reproducing an Aurora classic added up to a reasonable business gamble.

The next problem was merchandising. At the time Playing Mantis had no distribution to hobby shops, and the big chain stores seemed unlikely to carry the Haunted House. Then a representative from F. A. O. Schwarz stopped by the Playing Mantis showroom and happened to spy a copy of the Haunted House. He arranged for Schwarz to list it in their Fall 1995 Christmas catalog as a new product by a new company: Polar Lights. The word play in the company name and the familiar oval logo announced that Aurora was back in spirit. Hobbyists were delighted! The $79 price tag hurt, but it was much better than $500, and Schwarz disposed of the first run of kits quickly. Then in 1996 the Haunted House appeared in stores and hobby shops at $17.99. Polar Lights announced that more kits would soon be forthcoming. For Aurora fans, a new dawn was breaking.

The Mummy's Chariot rolled into the F. A. O. Schwarz catalog in 1996 and on to store shelves that same year. The versions of the Haunted House and Mummy's Chariot kits released on the general market featured a new highlight: Frightening Lightening glow plastic parts. The next Aurora-inspired kits didn't hit the stores until 1997, but they were all long-lost classics: the Bride of Frankenstein, Frankenstein's Flivver, and the Munsters living room scene.

Plastic model connoisseurs loved the re-release choices. All of the Polar Lights selections were models that had not been great sellers for Aurora the first time around, and thus were difficult or impossible to find in on the collector's market. Clearly, Polar Lights was aiming at a target audience of hard-core adult male Auroraphiles, not the general public. Tom Lowe explained simply that he was bringing back the kits he liked best. (Actually, in these early days the selection of new kits sometimes depended on which kits could be found on the collector market.) Beyond that, Polar Lights aimed for a company image pitched to a high level of exuberance: "Unusually Fun Model Kits!" As Lowe put it, "If it isn't cool, we won't do it."

Kit builders had only one complaint about the new models: When they tried to assemble the kits, they found that ordinary plastic cement from a tube would not hold the parts together. It turned out that the ABS plastic used by the Chinese manufacturer required adhesives with stronger solvents, such as super glue. The staff at Polar Lights, novices in plastic models, had not anticipated this problem. The short-term response was to place a note in each kit box listing common cements that would work on ABS. After that, Polar Lights switched to styrene plastic in all subsequent releases.

Between 1997 and 1999 Polar Lights introduced almost two dozen Aurora kits to the public. From a business point of view, this may have been too much of a good thing. "We saturated the market," concluded Lowe. Some of the new kits did not sell as briskly as the earlier ones. On the other hand, Polar Lights started introducing

Thanks to Polar Lights, in 1999 the grandchildren of Aurora's original customers could again experience the thrill of Frankenstein, Dracula, the Wolf Man, and the Mummy—

some of its own original models that sold very well. The Jupiter II from Lost in Space satisfied a long-held craving by many sci-fi fans.

Any company aiming to follow in the Aurora tradition would have to expect some controversies. For example, Lowe knew that the Guillotine had stirred up a storm of protests back in the 1960s because of its unusual working feature. He wanted to stay clear of anything that suggested blood and gore, rather than fun. Yet, Polar Lights fans, who could contact the company by its on-line computer bulletin board, repeatedly clamored for the return of the guillotine. After doing some research on psychological implications (like Bill Silverstein some forty years earlier), Lowe relented and the Guillotine made a comeback.

When Polar Lights reintroduced the Hunchback on Notre Dame, it used the name "Bellringer" out of deference to Disney's use of the name in its recent movie. However, the same approach did not work for Godzilla's Go Cart because simply leaving out the "Godzilla" name and airbrushing Godzilla's distinctive dorsal fins out of the picture on the box did not appease the masters of Japan's beloved monster. They asked that distribution of the kit be stopped and the remaining inventory of kits destroyed. Polar Lights complied.

When a new *Planet of the Apes* movie appeared in 2001, Polar Lights presented hobbyists with four of the figure kits that had originally been produced by Addar in the 1970s. At the request of Toys "R" Us, the kit boxes bore the familiar Aurora oval logo. Thus, in an interesting twist of history, a set of models originally produced by men exiled from Aurora, finally appeared under the Aurora trademark.

By 2003 Polar Lights had brought back so many of the most sought-after Aurora models that it was becoming difficult to decide upon another kit release that might attract enough buyers to make it a worthwhile business investment. All-in-all, Polar Lights had done remarkable work in recovering lost Aurora treasures.

Revell-Monogram-Aurora

In 1986 Monogram and Revell had been purchased by a group of investors and the operations of the two companies were consolidated at the Monogram plant in Morton Grove. Soon models that had been developed by one company appeared in stores under the brand name of the other company. Figure kit modelers received a pleasant surprise in 1999 when some old Aurora stalwarts came back in boxes with the Revell name. Superman, Batman, and Robin—the most famous of the superheroes—made a nice trio of models for comic book fans. Then, in a real surprise, the Monsters of the Movies versions of Frankenstein's monster and Dracula were released. (So the molds did survive!) When Polar Lights released its Chinese-made copies of Rodan and Ghidorah the next year, half the Monsters of the Movies series was once again available to builders.

As the new millennium began it appeared that some of Aurora's models had achieved the status of timeless classics. Both old-timers and youngsters can look forward to seeing more from Revell-Monogram and Polar Lights in the way of retro-recreations in the Aurora tradition.

Today a shopper visiting a hobby shop or toy store may, if he knows what to look for, find several kits produced from Aurora molds. However, to locate original Aurora products the search must be directed toward toy and kit collector publications, hobby clubs, and internet auction sites. Although Aurora kits are among the most sought after models, many of the basic Aurora kits—the World War I biplanes, for example—are readily available at prices under $20. The selling prices for figure and science fiction kits went through the roof in the 1990s but have receded in recent years.

Some rare models are difficult to find at any price. To fill this void some small "garage" companies have issued resin recasts of obscure kits with a market only among extreme aficionados. It is interesting that a company which began in a garage has prompted the rise of a new generation of garage companies.

When Aurora shut down its molding machines in the summer of 1977 there were few who lamented its demise, for the company seemed to have outlived its usefulness. It was remembered as a pioneer in introducing plastic modeling to the public, but criticized as a company that had failed to keep up with the times.

However, something was lost with the passing of Aurora. Generations of kids growing up in the 1950s, 1960s, and 1970s derived immense satisfaction from building Aurora's simple, colorful models. Aurora's kits may not have been the best, but building them was the most fun. The more sophisticated model companies of today have not succeeded in capturing the hands, hearts, and imaginations of youngsters quite the way Aurora did with its biplanes, jets, knights, monsters, hot rods, ships, tanks, comic book heroes, sports cars, and science fiction creations.

Illustrated Directory of Aurora Plastic Kits

Models in this index have been categorized by subject and further subdivided into categories of kits which form logical units. This arrangement should be useful for most collectors, but not all kits will appear in the sequence of their catalog number.

The first notation on each kit is its catalog number. Other catalog numbers under which it appeared are also indicated. The second piece of information is the kit's name. The dates indicated for each kit are those for the years it appeared in the United States catalog. Often a kit was introduced the year before it appeared in the catalog, and a few times kits were released after the year they made the catalog. A few rare kits did not make the catalog at all. Some kits were produced in Canada, Holland, or England after they no longer appeared in the US catalog. A few kits were produced only in Canada or England.

The scales indicated, with a few exceptions, are those appearing in John Burns' *Collectors Value Guide* and are used with his permission. The precise accuracy of scales given may be open to question, but they are useful in distinguishing one kit or kit series from another.

The plastic colors indicated for each kit are those known to exist. Aurora had a fixed color for each kit, but changing any model's color was as simple as dumping in a different color dye along with the next batch of colorless raw plastic pellets being fed into the molding machine.

The values included in this guide are intended to give collectors a general idea of the prices being paid for each kit. However, the price of any given kit will vary widely depending on supply and demand.

The wide variance in the prices indicated for some of the kits listed can be explained by two factors: the age of the kit and packaging. The age of a kit is very important in determining value. An older issue is almost always more highly valued than a newer one. Kits made in the United States bear a copyright date, but that date indicates the original date the mold or box art was created, not necessarily the year a particular copy of that kit was produced.

Packaging variations are the most commonly used way of dating the age of a kit, and by far the most important variable is the Aurora logo that appears on the box. Here is a summary of the evolution of Aurora's trademark:

1952-54 The first two kits, **22, 33** were the only ones to appear in gray, one-piece boxes. All others are in full color boxes, with a monochrome Aurora rising sun logo on the box end and the "Famous Fighters" trademark name on the box top. Most kits with a Brooklyn address box have "U-Ma-Kit" on the box sides.

1955-56 Boxes have "Famous Fighters" and a small rectangular sunburst logo reading "The Aurora Line" on the box top and the one color rising sun trademark on the box end.

1956-57 "Famous Fighters" still appears on the box top, but in smaller lettering. A large, more colorful rectangular sunburst logo reading "Famous Fighters by Aurora" appears at the corner of the box top and on the box ends.

1957-64 The oval logo with "Famous Fighters" in its borders appears on both the box top and ends. The writing in the oval logo borders will be different for some kits: "Guys and Gals," "Famous Sports Cars," "Playcraft Hobby Kits," etc.

1963-75 "Famous Fighters" is dropped from the borders of the oval logo.

1969-72 The large "A" logo is adopted at the same time as the switch to white, square boxes. Some kits are not repackaged and continue the flat rectangular box, oval logo style down to 1975.

1970-77 AURORA in block lettering appears on some boxes in 1970 and on all boxes after 1975.

The price marked on a kit's box end is also an indication of age. The digits to the right of the hyphen in a kit number are the company's recommended sale price. Thus kit number 101-69 is a .69 cent kit. Since prices rose over the years, a 101-69 kit is older than a 101-79 issue or a 101-100 copy.

Organization of Kit Directory

Aircraft

World War I

All original issue World War I aircraft have auxiliary parts molded in black plastic and a ground base with wheel chocks and at least one ground crew figure. All original issues have raised decal locater lines.

100 Sopwith Triplane: 1964-70, 1/48; **$25-35**
Molded in black plastic to represent Raymond Collishaw's "Black Maria" of the Royal Navy's "Black Flight." Pilot figure is climbing into the cockpit, while second pilot walks across base. Reissued by K & B in 1972 as **1100**. Box art by Kotula.

101 Nieuport 11: 1956-75, 1/48; **$15-40**
Olive plastic until 1974, when it became silver. Seated pilot and mechanic set to spin prop. Model bears the Indian head insignia of France's Lafayette Escadrille. The Le Prieur rockets on the wing struts are over scale. Reissued in 1977 as **754**. First box art by Cox. Second box art by Steel. Reissued by Allan Brookers of New Zealand in 1980 in Monogram box. Mold copied by Merit of England in the 1950s; this mold was used by Artiplast of Italy and SMER of Czechoslovakia.

102 Sopwith Camel: 1956-75, 1/48; **$15-40**
Olive plastic until 1974, when it becomes tan. Seated pilot and mechanic oiling wheel. Carries four bombs. The large cut-out in the wing above the pilot was unusual in Camels. Aircraft markings are for Roy Brown's plane. Revised and reissued in 1976 as **751**. First box art by Cox. Second box art by Steel. Reissued by Monogram. Copied by Merit, reissued by Artiplast and SMER.

103 SE-5: 1956-75, 1/46; **$15-40**
Olive plastic until 1974, when it became tan. Seated pilot with right arm raised, mechanic working on engine with wrench. The nose of this model more closely resembles the SE-5A. Markings are for James McCudden's plane. Reissued in 1977 as **SE-5A 755**. First box art by Cox. Second box art by Kotula. Yellow box art by unknown artist. Reissued with modifications by Monogram and by Allan Brookers of New Zealand in 1980 in Monogram box. Copied by Merit, reissued by Artiplast and SMER. A second copy of the mold was made by Marusan of Japan in late 1950s. This Marusan copy was also issued by OM, Fuji, Entex, and Sunny.

104 Albatross D-3: 1956-75, 1/46; $15-40

Dark green plastic until 1974, when it becomes matte red. Seated pilot and mechanic set to spin prop. The shape of the fuselage is inaccurate and there are dual radiators on the upper wing, not the usual single one. "Albatros" is misspelled. Reissued with revisions in 1976 as **752 Albatros D-5**. First box art by Cox. Second box art by Steel. Reissued by Allan Brookers of New Zealand in 1980 in Monogram box. Copied by Merit, reissued by Artiplast and SMER. Also copied by Marusan of Japan.

105 Fokker Dr-1: 1956-75, 1/43; $15-40

Metallic burgundy plastic until 1974, when it becomes matte red. Seated pilot and mechanic steadying the wing. Box art changed in 1964 to include a "news clipping" about Richthofen's death. The model aircraft's serial number 2009/17 was a Fokker factory number which did not appear on the side of the actual Richthofen aircraft. Reissued in 1976 as **750**. Box art by Cox. Copied by Merit, reissued by Artiplast and SMER.

106 Fokker D-7: 1956-75, 1/46; $15-40

Dark metallic green plastic until 1974, when it becomes matte green. Seated pilot with head turned to right and mechanic pulling out a wheel chock. Reissued in 1976 as **753**. Box art by Cox. Reissued by Monogram and by Allan Brookers in Monogram box. Copied by Merit, reissued by Artiplast and SMER

107 Spad XIII: 1958-66, 1/48; $50-70

Olive plastic. Seated pilot and standing soldier with briefcase. This is the rarest of the World War I kits today because it sold well and was never reissued by Aurora. Markings for American 22nd Aero Squadron. Box art by Kotula. Reissued by Glencoe.

108 Nieuport 28: 1958-66, 1/48; $20-40

Light gray plastic. Seated pilot and mechanic lifting tail. Decals represent the American "Hat in the Ring" Squadron. Reissued by K & B in 1972 as **1108**. Box art by Kotula. Renwal based its kit on this model. Reissued by Glencoe.

109 Pfalz D-3: 1958-66, 1/48; $20-40

Gray plastic. Seated pilot and mechanic standing on wing next to pilot. Based on drawings by William Wylam in *Model Airplane News*. Reissued by K & B in 1972 as **1109**. Box art by Kotula. Reissued by Glencoe.

109-650 Dogfight: 1975, 1/48; $50-60

This is a Young Model Builders Club issue in a corrugated mailing box with a paste-on paper label of the "news clipping" Fokker Triplane box art and the John Steel Sopwith Camel art. The kits inside are the **750** Triplane and **751** Camel. A single instruction sheet covers both models and explains that the aircraft represent those of Manfred von Richthofen and Roy Brown. Box art by Cox/Steel.

112 DeHavilland DH-4: 1958-66, 1/48; $35-50

Olive plastic. Seated pilot points upward, mechanic looks up through binoculars, gunner stands in rear cockpit. Based on drawings by William Wylam in *Model Airplane News*. An accurate model, never reissued by Aurora. Box art by Kotula.

113 Bristol F2B: 1958-66, 1/48 $35-50

Olive plastic. Two seated crewmen and mechanic with arms raised. Based on drawings by William Wylam in *Model Airplane News*. Reissued in 1976 as **776**. First box art by Kotula. Second box art by Steel.

114 JN-4D Jenny: 1958-64, 1/48; $30-50

Yellow plastic. Seated pilot, standing mechanic, kneeling mechanic. Decals for a trainer based at Kelly Field. A comparison of this kit to Lindberg's Jenny shows Aurora's typically "heavy" details. Box art by Kotula.

125 DeHaviland DH-10: 1958-62, 1/48; **$35-60**
Olive plastic. Three crew figures, kneeling mechanic, mechanic reaching up with wrench. "DeHavilland" is misspelled on the box. Based on drawings in *Aircraft of the 1914-1918 War*. Reissued in 1972 as K & B **1125** and in 1977 as **786**. Box art by Kotula.

126 Gotha: 1958-62, 1/48; **$35-65**
Metallic burgundy plastic. Three crew figures, two mechanics with arms raised. Eight externally carried bombs. In 1971, hobbyist Ross Abare ordered a run of Gotha's from Aurora which are the most common editions found today. The Abare issue comes in a box with a zip code after the address, does not have the price in a circle on the box top, and its logo does not read "Famous Fighters." However, the only difference between it and the last authentic 1960s issue is that the manufacturer's code on the back of the decal sheet includes the number "71," disclosing the date of printing. Also reissued in 1972 as K & B **1126** and in 1977 as **785**. Box art by Kotula.

134 Fokker E-3: 1964-69, 1/42; **$25-40**
Tan plastic. Seated pilot gestures with left hand while mechanic holds on to the wing tip. Made in larger scale than 1/48 to fit into a standard box. Reissued in 1972 as K & B **1134**. Box art by Kotula.

135 Fokker D-8: 1960-66, 1/48; **$40-50**
Metallic green plastic. Seated pilot and mechanic set to spin prop. Reissued in 1972 as K & B Fokker E-5 **1135**. Box art by Kotula.

136 Halberstadt CL-II: 1960-69; 1/48 **$50-60**
Gray plastic. Seated pilot, mechanic hands up notebook to gunner in rear cockpit. Reissued in 1972 as K & B **1136**. Box art by Kotula. Mold destroyed in train wreck on way to Monogram.

141 Breguet 14: 1964-70, 1/48; **$40-50**
Green plastic. No crew figures, two mechanics push on lower wings. Lacks bomb rack and bombs. Reissued in 1972 as K & B **1141**. Box art by Kotula. Mold destroyed in train wreck on way to Monogram.

142 Albatross: 1964-70, 1/48; **$60-70**
Gray plastic. Standing pilot hands equipment up to gunner in rear cockpit. A C-III observation plane. Nicely detailed model. "Albatros" is misspelled. Reissued in 1972 as K & B **1142**. Box art by Kotula. Mold destroyed in train wreck on way to Monogram.

200 Barnstormers Customizing Kit: 1960, 1/48; **$90-110**
Contains a yellow DH-4 **(112)** and a green JN-4D **(114)**. Although the plastic colors were changed, the parts are the same as the original issues. The flamboyant "barnstormer" decals do not match the raised decal lines on the plastic.

201 Great Air Battles of World War I: 1962, 1/48; **$150-180**
Issued in cooperation with *American Heritage* magazine. Set contains a Nieuport 11 **(101)** and Fokker Dr-1 **(104)**. Box says it contains a booklet on historic air battles, but no existing kits have this booklet. Box art by Leynnwood.

398 Fokker D-7: 1968-75, 1/19; **$85-95**
Red, black, and brown plastic, metal screws. Originally a flying model, converted to a "screwdriver" kit. The decals are early-war iron crosses, not the Latin crosses used on actual D-7s. Box art by Steel.

399 SE-5: 1968-75, 1/19; **$85-95**
Military green, black, and brown plastic, metal screws, rubber tires. Originally a flying model, converted to a "screwdriver" kit. Box art by Steel.

700 Series. In 1976-77 Aurora reissued nine of its World War I models in white boxes with photographs of built models on the box. The ground bases and mechanics were eliminated, and the parts formerly molded in black were incorporated into the main molds. Fabric areas of the aircraft were given a texture to simulate cloth. Raised decal locater lines were removed. Two sets of new decals were included with each kit.

750 Fokker Dr-1: 1976-77, 1/43; **$12-15**
Red plastic. Skids added to wing tips. Revised version of **105**.

751 Sopwith Camel: 1976-77, 1/48; **$12-15**
Olive plastic. Cockpit opening changed and large cutout in upper wing narrowed. Bombs deleted. Revised version of **102**.

752 Albatros D-5: 1976-77, 1/48; **$12-15**
Tan plastic. Altered from D-3 by changing the shape of the tail and altering fuselage detail. Second radiator removed from upper wing. Includes a paper pattern for painting lozenge camouflage on wings. Revised version of **104**.

753 Fokker D-7: 1976-77, 1/46; **$12-15**
Blue plastic. Pilot's step and hand loops added to fuselage, but tail skid left out. Interplane struts thinned to proper scale. Revised version of **106**.

754 Nieuport 11: 1977, 1/48; **$12-15**
Silver plastic. Still retained black parts. Revised version of **101**.

755 Airco SE-5A: 1977, 1/46; **12-15**
Olive plastic. Retained black parts. Revised version of **103**. Designation changed from SE-5 to SE-5A since shape of nose conforms to A variant.

776 Bristol F2B: 1976-77, 1/48; **$20-25**
Tan plastic. Revised version of **113**.

785 Gotha G-V: 1977, 1/48; **$25-30**
Green plastic. Revised version of **126**.

786 DeHavilland DH-10: 1977, 1/48; **$25-30**
Olive plastic. Revised version of **125**.

Between the Wars Aircraft

110 Tiger Moth: 1959-66, 1/48; **$30-40**
Olive, clear plastic. This mold was developed by Toltoys of Australia and sold to Aurora. The model's detailing is inferior, but the kit does contain an assembly jig to help set the wings at the proper dihedral. Copied by Merit of Great Britain. Box art by Kotula.

111 M-2 Mailplane: 1958-64, 1/48; **$25-35**
Burgundy, black, clear plastic. Includes a base and standing pilot figure like the World War I kits. Douglas biplane with decals for Los Angeles based "Western Air Express." Reissued in 1976 as **775**. Box art by Kotula.

115 Boeing P-26A: 1956-59, 1/43; **$35-40**
Turquoise, clear plastic. Movable flight controls. Decals for Army's 34th Pursuit Squadron with "Thunderbird" insignia. Reissued in 1972 as K & B **1115**.

116 Curtiss P-6E Hawk: 1956-70, 1/44; **$35-40**
Dark green, clear plastic. Movable flight controls. Decals for Army's 17th Pursuit Squadron.

117 Curtiss SBC-3 Helldiver: 1956-69, 1/41; **$40-50**
Silver, clear plastic. One of the more detailed early kits. It has movable flight control surfaces and good interior cockpit detail for a 1950s kit. Heavy rivets, raised decal lines. Two crew figures, two underwing bombs, belly fuel tank.

121 Boeing P-12E: 1958-66, 1/48; **$35-40**
Olive, clear plastic. Has ground base, seated pilot and standing mechanic like the World War I kits. Decals for Army's 27th Pursuit Squadron with "Diving Hawk" insignia. Box art by Kotula.

122 Boeing F4B-4: 1958-66, 1/48; **$25-35**
Gray and clear plastic. Includes ground base, seated pilot and standing mechanic like World War I kits. Two bombs under wings. An unusual 1950s kit in that it does not have decal locater lines. Decals for Navy's Fighting Squadron One-B with "High Hat" insignia. Reissued in 1972 as K & B **1122**. Box art by Kotula. Copied by Marusan of Japan in the 1950s; this copy reissued by Entex in 1975.

775 M-2 Mailplane: 1976-77, 1/48; **$20-25**
Red plastic. Revised version of **111**. Surface textured to simulate fabric and raised decal lines removed. No clear or black parts, no ground base or mechanic as in original issue. Because auxiliary mold for the black parts had been lost, Aurora made a new propeller and wheels. New Indian Head markings on decal sheet.

World War II Aircraft

20 Spitfire: 1954-74, 1/43; **$25-110**
Light blue plastic in first issue, then light metallic blue. First issue had wheels, bombs, and raised decal lines, but no rivets. Rivets added to later editions. The number "9" inscribed on the left fuselage is reversed. Reissued as **342**. Airfix's first kit was a scaled-down copy of Aurora's model—right down to the blue plastic and inaccurate unit ID letters on the fuselage. First box art by Cox. Second box art by Künstler.

30 Focke Wulf: 190 1954-74, 1/47; **$30-110**
Black plastic. Raised decal lines. The ID number that appears on the tail is one painted on a captured FW-190 by the US. Earliest edition has swastika on tail fin; replaced in later issues with a cross. In 1961, each wing was divided into an upper and lower part to make the parts thinner, thus speeding-up cooling time in the molding machine and eliminating "sink" spots often found in thick parts caused by uneven cooling. Reissued as **344**. First box art by Cox. Second box art by Künstler.

40 F6F Hellcat: 1954-69, 1/51; **$35-110**
Metallic blue plastic. Raised decal lines. Originally each wing was molded in one piece. About 1960 each wing was divided into an upper and lower part. First box art by Cox. Second box art by Künstler.

44 P-40 Flying Tiger: 1953-74, 1/48; **$25-150**
Aurora's first original kit. A P-40E. First issue was gray plastic with early World War II insignia. Later issues switch to olive and sometimes silver plastic. Raised decal lines. First issue in Brooklyn box. Landing gear, rivets, and six bombs added in 1954. Wings split into upper and lower parts in 1960. Box art for first issues by Cox. Next by Kotula. Final box art by Steel.

55 Messerschmitt ME-109: 1953-74, 1/46; **$45-160**
First issue is light red plastic, changed to metallic burgundy in 1954, and to gray or light blue in the 1970s. First issue in Brooklyn box. Landing gear, rivets, heavier decal lines, and bombs added in 1954. First issue has swastika on tail fin; replaced by a cross in later editions. The "c" is omitted from the spelling of "Messerschmitt" on the first issue box. First issue box reads "U-Ma-Kit" on side. Reissued as **343**. First of several box illustrations by Cox. Final box art by Künstler.

70 AT-6 Texan: 1956-69, 1/48; **$15-65**
Silver, clear plastic. Has no inscribed decal locater lines because it was also issued in Navy markings as **80**. This was the first Aurora model to have a separate pilot figure and cockpit opening in the fuselage. Second box art by Künstler.

80 SNJ Trainer: 1956-70, 1/48; **$25-65**
Yellow, clear plastic. Has no inscribed decal locater lines because it was also issued in Air Force markings as **70**. HMS's designers gave the SNJ decal markings for a Navy Reserve plane stationed at HMS's home, Willow Grove, Pennsylvania. First box art by Ed Marinelli. Second box art by Künstler.

81 P-47 Thunderbolt (Co): 1965-69, 1/53; **$25-35**
Ex-Comet. Olive, clear plastic. A Republic P-40D with bubble canopy. Includes two underwing tanks, but no landing gear. Comet teardrop stand. Reissued as **341**. Box art by Leynnwood.

88 Jap Zero: 1953-70, 1/48; **$50-160**
Yellow, clear plastic. Raised decal lines. First issue in Brooklyn box. Bombs, rivets, and landing gear added in 1954. Wing parts split into upper and lower halves in about 1960. Name changed to "Japanese" Zero in 1958. Early box illustrations by Cox. 1960 box art by Künstler.

99 P-38 Lightning: 1953-74, 1/48; **$40-100**
Dark blue plastic in first issue, then light metallic blue. Later issues in olive or silver. Some models issued in the late 1950s were chrome plated. Raised decal lines. A P-38L. Last of the kits to be released in a Brooklyn box, and last model originally issued without landing gear and rivets; these were added in 1954. Reissued as **346**. First box art by Cox. Second box art by Kotula.

118 P-51 Mustang: 1957-74, 1/48; **$40-90**
Early issues olive plastic, later silver. A P-51H. First Aurora kit with retractable landing gear; also removable engine access panels. Otherwise an ordinary model with raised decal lines and heavy rivets. Reissued as **345**.

286 P-47 Thunderbolt (Co): 1965-68, 1/88; **$15-20**
Ex-Comet. Olive, clear plastic. A smaller scale version of **81** without underwing fuel tanks. Includes teardrop Comet stand. Used the same box art as **81**. Box art by Leynnwood.

341 P-47 Thunderbolt (Co): 1965, 1/53; **$140-160**
Reissue of **81**. Ex-Comet. Part of the "Twelve O'Clock High" set, with regular box art and decals, but with a band on one end of the box identifying it as a kit issued in connection with the mid-60s TV series. All kits in this series were molded in olive plastic. Box art by Leynnwood.

342 Spitfire: 1965, 1/43; **$150-170**
Reissue of **20**. Olive plastic. Twelve O'Clock High set.

343 Messerschmitt ME-109: 1965, 1/48; **$150-170**
Reissue of **55**. Olive plastic, but some regular metallic burgundy plastic kits were packaged as part of the Twelve O'Clock High set.

344 Focke Wulf FW-190: 1965, 1/48; **$150-170**
Reissue of **30**. Olive plastic, but some are regular black plastic. Twelve O'Clock High set.

345 P-51 Mustang: 1965, 1/48; **$150-170**
Reissue of **118**. Olive plastic. Twelve O'Clock High set.

346 P-38 Lightning: 1965, 1/48; **$150-170**
Reissue of **99**. Olive plastic. Twelve O'Clock High set. Box art by Steel used only in this set.

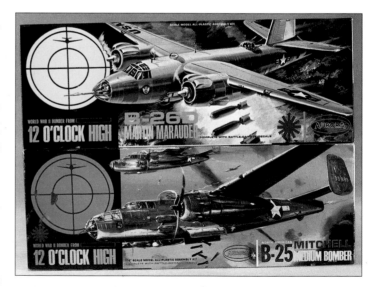

347 B-26 Marauder: 1965, 1/46; **$200-220**
Reissue of **371**. Olive plastic. Twelve O'Clock High set. Special "Battle Damage" decal sheet. Box art by Künstler.

348 B-25 Mitchell: 1965, 1/45; **$200-220**
Reissue of **373**. Olive plastic. Twelve O'Clock High set. Special "Battle Damage" decal sheet. Decal insignia for the 5th Air Force of the South Pacific theater are changed to European 8th AAF logos since the TV series was set in Europe. Raised decal locater lines on the mold were not changed. Box art by Steel.

352 B17 Bomber Formation: 1965-70, 1/156; **$350-375**
Olive plastic. Decals with "battle damage." Color instructions and color print of box art for framing. Twelve O'Clock High set. A diorama of three B-17's over a black plastic landscape base of a city with factories, airfield, and railroads. Clear stands have explosions molded-in at points of attachment and decals of strings of dropping bombs. Reissue of **491** aircraft. Box art by Künstler.

377 P-51 Mustang (Co): 1969-75, 1/27; **$90-100**
Dark blue wings, light blue fuselage, black propeller. Clear canopy. A flying model purchased from Comet, it was never issued as a flying model by Aurora, but it was converted into a "screwdriver" assembly static model.

378 Curtis P-40 (Co): 1969-75, 1/27; **$90-100**
Dark olive fuselage, black propeller, gray plastic accessories and underside wing parts. Converted from Comet flying model into "screwdriver" model.

371 Martin B-26 Marauder: 1954-75, 1/46; **$45-100**
Silver, clear plastic. Rivets, raised decal lines. Nose decal of 8th Air Force emblem, which planes never carried, and "USAF" wing markings that didn't appear until after World War II! A simple model, it was among Aurora's best sellers for years. Also issued as **347**. Final box art by Schaare.

372 Boeing B-29 Superfortress: 1955-75, 1/76; **$50-120**
Silver, clear plastic. Rivets, raised decal lines. Incorrect nose art of 20th Air Force emblem and anachronistic "USAF" wing decals. Turrets rotate. Final box art by Schaare.

373 North American B-25 Mitchell: 1955-75, 1/46; **$50-110**
Silver, clear plastic. Rivets, raised decal lines. Incorrect nose art of 5th Air Force and anachronistic "USAF" wing decals. A B-25J. Also issued as **348**. Second box art by Steel.

374 PBY Catalina: 1956-69, 1/74; **$55-90**
Silver, clear plastic. Rivets, raised decal lines. Wing-tip pontoons can be raised and lowered. Reissued in 1973-75 as **396**. Second box art by Leynnwood.

392 P-61 Black Widow: 1960-77, 1/48; **$30-70**
Black, clear plastic. Northrop P-61C. Raised decal lines, but no rivets. No wheel wells for landing gear. First box art by Kotula.

396 PBY-5 Catalina 1973-75, 1/74; **$35-40**
Silver, clear plastic. Issued as **374** from 1956-69.

397 B-25 Mitchell Bomber: 1969-75, 1/32; **$180-200**
Olive, black, silver, clear plastic. Rubber tires. Originally issued as a flying model in 1960. Modified to be assembled with screws as "screwdriver" kit. Original flat box, oval logo issue is more valuable than later square box issue. Box art by Künstler.

491 Boeing B-17 Flying Fortress: 1957-71, 1/156; **$50-60**
Olive, clear plastic. A B-17G. The small scale of this model made fine detail impossible. Reissued as part of **352** Twelve Clock High diorama. Box art by Kotula.

497 F4F Wildcat: 1962-71, 1/65; **$15-20**
Dark blue, clear plastic. Includes two underwing fuel tanks and landing gear. Box art by Steel.

498 P-38 Lightning: 1962-71, 1/84; **$15-20**
Dark blue, clear plastic. Netherlands issue in olive plastic. Box art by Leynnwood.

GB-798 Giant Bomber Gift Set: 1955-60; **$350-400**
Includes **371, 372, 373.**

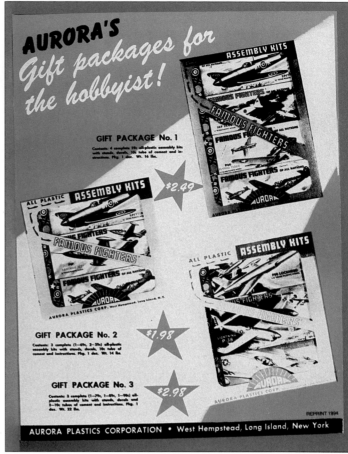

1954 Gift Packages
The 1954 gift sets were packaged in thin cardboard sleeves that held three or four regularly-boxed kits together. Presumably any combination of kits in the same size box could be packaged together. A tube of glue was included. None of these gift packages are known to have survived.

Gift Package No. 1: 1954
The 1954 catalog shows the following four kits in the package: 20 Spitfire, 88 Zero, 44 Flying Tiger, and 55 Messerschmitt. It sold for $2.49.

Gift Package No. 2: 1954
The 1954 catalog shows the following three kits in the package: 20 Spitfire, 22 Panther Jet, 55 Messerschmitt. It sold for $1.98.

Gift Package No. 3: 1954
The 1954 catalog shows the following three large-box kits: 33 F-90, 99 P-38, 1977 F-86D. It sold for $2.98.

Post-War Military Aircraft

22 F9F Panther: 1952-74, 1/49; **$25-160**
First issue dark blue plastic; later issues blue metallic plastic. Aurora's first kit, it was copied from a Hawk model. First issued without landing gear, rivets or rockets, which were added in 1954. Earliest box is one-piece gray cardboard with "U-Ma-Kit" on the sides. Second issue box still carries Brooklyn address. In production for 22 years, the longest run of any Aurora kit. First color box art by Cox. Second box art by Kotula.

33 F-90 1952-70, 1/48; **$55-175**
First issue gray plastic; later issues silver plastic. Issued with the **22** F9F in the fall of 1952. Copied from Hawk. First issued without landing gear, rivets, or rockets, which were added in 1954. Earliest box is one-piece gray cardboard with "U-Ma-Kit" on the sides. Second issue box still carries Brooklyn address. First box art versions by Cox. 1959 issue box art by Kotula.

50 Lockheed XFV-1 "V. T. O." 1954-60, 1/48; **$140-160**
Silver, green tinted clear plastic. Early .79 and .89 cent editions come in slightly smaller box, with large "Famous Fighters" logo on box top, and "VTO"—not "50"—on box end. Later .89 cent editions come in larger box with "Aurora Line" logo on box top. Heavy riveting and raised decal lines. Twin propellers geared to spin in opposite directions.

60 Convair V. R. Vertical Riser XFY-1 "Pogo" 1954-59, 1/48; **$80-110**
Silver, green tinted clear plastic. Twin propellers geared to spin in opposite directions.

66 YAK-25/MIG-19 1953-70, 1/48; **$70-200**
Green plastic in first issue, changed to metallic green in later issues. Originally issued as the Yak-25, the name was changed to MIG-19 in 1954 when landing gear, rivets and eight underwing rockets were added. First box has Brooklyn address. Earliest MIG-19 issue is most valued. Yak box art by Cox. Silver Mig box art by Kotula.

1977 F-86D Sabre Jet: 1953-70, 1/48; **$25-100**
Silver, clear plastic. First edition kit in Brooklyn box. In 1954 a heavy encrustation of rivets was added to this model of an aircraft noted for flush riveting! Also added: landing gear and eight under-wing rockets, which the actual aircraft never carried! First box art versions by Cox. 1959 issue box art by Kotula.

82 F-104 Starfighter (Co): 1965-69, 1/62; **$15-20**
Ex-Comet. Gray, clear plastic. A Lockheed F-104A. No landing gear. Includes teardrop Comet stand. Issued in England in 1973.

119 F8U Crusader: 1957-74, 1/50; **$20-60**
Issued in gray or white, clear plastic. Features retractable landing gear and tail section that pulls off so jet engine can be removed. This was Aurora's response to Monogram's models with moving parts. Based on the aircraft's prototype, the model differs considerably from the real plane. Box art with plane on ground by Ed Marinelli.

120 X-15 Satelloid Rocket Plane: 1959-74, 1/48; **$100-160**
First .98 cent issue is in white plastic and is most valuable. White—because earlier research aircraft had been white, and Aurora had not seen the actual black X-15. Only eleven parts to this model. Raised decal lines, no cockpit opening, and no landing gear. Second black plastic issue corrects the markings to match the real aircraft: the "X-15" logo is moved from the tail fin to the nose and the serial number "66670" is put on the tail fin. Landing gear is added, the exhaust ring is made larger, and an instrument boom is placed into the nose. NASA decals added in mid-1960s. First box art by Kotula. Second box art by Leynnwood.

123 F-105 Thunderchief: 1959-70, 1/78; **$20-30**
Silver and clear plastic. Instrument boom in nose as in FH-105 prototype. Model includes four underwing rockets and removable belly weapons pod. Wheel well openings in wings, but not fuselage. Box art by Kotula.

124 Avro Arrow CF-105 1959-64, 1/80; **$35-45**
White, clear plastic. Model of a Canadian fighter aircraft. *Scale Modeler* (December 1975) says: "remarkably accurate model in outline shape, once the typical heavy Aurora rivets and lines had been sanded off." Box art by Kotula.

128 Nuclear Powered Bomber: 1959-64, 1/182; **$50-70**
Light gray, clear plastic. A simple model with little surface detail, raised decal lines, and no landing gear wells. Russian aircraft issued with Russian Airliner **127**. Model is based on an article in *Aviation Week* (December 1, 1958). The actual aircraft was the conventionally powered "Bounder." Box art by Kotula.

137 CF-100 1960-69, 1/67; **$25-35**
White, clear plastic. Canadian fighter. Fuselage divided into top and bottom halves, not right and left halves. No wheel wells for landing gear. Box art by Kotula.

138 Cessna T-37 (St): 1962-74, 1/43; **$20-30**
Light gray, clear plastic. Rivets, but no decal locater lines. Has landing gear wells. Reissued as **147 Cessna A-37**. Mold was purchased from Strombecker.

139 Temco TT-1 (St): 1962-69, 1/42; **$20-30**
Yellow, clear plastic. Rivets, but no decal locater lines. Has landing gear wells. Mold was purchased from Strombecker.

140 Northrop N-156 Freedom Fighter: 1962-74, 1/48; **$20-25**
White, clear plastic. Inscribed decal lines. Three underwing fuel tanks and two Sidewinder missiles. No wells for the landing gear. Box art by Leynnwood.

143 Boeing KC-135 (Co): 1965-70, 1/125; **$20-25**
Ex-Comet. Gray, clear plastic. Refueling boom raises and lowers. Box art by Leynnwood.

144 Convair B-58 Hustler (Co): 1965-69, 1/91; **$25-30**
Ex-Comet. Gray, clear plastic. Pilot figure, weapons pod, landing gear with wheel wells in wings only. No clear parts for side windows. Inscribed decal lines. Box art by Leynnwood.

145 S2F Hunter Killer (Co): 1965-69, 1/54; **$50-60**
Ex-Comet. Dark gray, clear plastic. This kit is in demand because it is one of the few models of an S2F. **288** is an S2F-1 in smaller scale. Box art by Leynnwood.

146 Hiller X-18 (Co): 1965-69, 1/70; **$60-80**
Ex-Comet. Silver, clear plastic. Model features working tilt-wings. Reissued in 1975 as **768** by the Young Model Builders Club with inscribed decal markings removed.

147 A-37 Strike Jet (St): 1968-70, 1/43; **$50-55**
Gray, clear plastic. Also issued as **138 T-37**. At height of Vietnam War this combat version of the Cessna trainer was released with new John Steel box art. Has six underwing bombs and two rocked pods not included in T-37 trainer kit. Mold purchased from Strombecker.

These cellophane-wrapped sets were designed for Christmas sales. They apparently contained randomly packaged models. All were advertised only in 1958. None are known to have survived.
200 2 Famous Fighters F-90 box art
201 3 Famous Fighters T-6 box art
202 3 Famous Fighters P-38 box art
205 4 Famous Fighters P-40 box art
205 4 Famous Fighters XFY-1 box art
211 Four .49 cent kits

4 Hobby Kits: 1978?; **$80-100**
This set has no number and appears in no catalog. It contains four models packaged in a large flimsy box with a label pasted on top. The instruction sheets are inexpensive reproductions of regular instructions, with all mention of Aurora removed. This set was created by the person who purchased the inventory of kits leftover when Aurora closed in 1977. Confirmed kits (sometimes incomplete) included in this set: **701 Forrestal** (with **721 Enterprise** decals!), **351 Boeing 727 Eastern**, **499 CH-54A Skycrane**, **568 1922 Ford Roadster**.

287 F3D Skyknight (Co): 1965-68, 1/97; **$15-20**
Ex-Comet. Gray, clear plastic. Decals for Korean War night fighter of Marine squadron VMF-542. Inscribed decal lines. No landing gear, 4 underwing rockets. Remained in Holland catalog until 1971. Also issued by Frog in 1956 and Addar in 1976.

288 S2F-1 (Co): 1965-68, 1/111; **$15-20**
Ex-Comet. Dark gray, clear plastic. Nicely detailed for a small scale kit. No landing gear. **145** is an S2F-1 in larger scale. Box art by Leynnwood.

289 North American F-100 Super Sabre (Co): 1965-68, 1/103; **$15-20**
Ex-Comet. Gray, clear plastic. Also issued by Frog in 1956 and Addar in 1976.

290 F-102 Dart: 1957-72, 1/121; **$20-25**
Silver, clear plastic. Model is of Convair prototype YF-102 with instrument boom in nose. Box art by Kotula.

291 Lockheed F-104 Starfighter: 1957-72, 1/96; **$15-20**
291 Lockheed F-104 Starfighter (Co) 1/110; **$15-20**
Silver, clear plastic. Both original Aurora kit and ex-Comet kit were boxed under number **291**. Both use the same decal sheet. The Comet mold contains a teardrop shaped Comet stand, not the triangular Aurora stand. The Aurora model fuselage is six inches long and the Comet five and a half. Only the Aurora model has landing gear.

292 Douglas F4D Skyray: 1957-70, 1/88; **$15-20**
Gray, clear plastic.

292 F4D Skyray (Co) 1/88; **$15-20**
Silver, clear plastic. Both original Aurora kit and ex-Comet kit were boxed under number **292**. The Comet mold contains a teardrop shaped Comet stand, not the triangular Aurora stand. The ex-Comet kit has a built-in pilot figure, which the Aurora does not. Only the Aurora model has landing gear. The ex-Comet kit was also issued by Frog in 1956 and by Addar in 1976.

293 F9F-6 Cougar: 1957-71, 1/80; **$15-20**
293 F9F-6 Cougar (Co) 1/80; **$15-20**
Gray and clear plastic. Both original Aurora kit and ex-Comet kit were boxed under number **293** with the same box art. The ex-Comet model lacks landing gear and has a pilot figure, which the Aurora model lacks. Otherwise the models are almost identical, except that the Aurora model has raised decal lines, while the Comet's decal locater markings are inscribed. Ex-Comet model also issued by Frog in 1956 and Addar in 1976. Box art by Ed Marinelli.

294 F-101 Voodoo: 1959-71, 1/136; **$15-20**
Silver, clear plastic. Box art by Kotula.

295 F-107 Fighter Bomber: 1959-70, 1/117; **$20-25**
Silver, clear plastic. Remained in British catalog until 1973. Box art by Kotula.

296 KC-135 (Co): 1965-68, 1/300; **$15-20**
Ex-Comet. Gray plastic. No clear parts, no landing gear. Inscribed decal lines. Also issued by Addar in 1976.

297 B-58 Hustler (Co): 1965-68, 1/175; **$20-25**
Ex-Comet. Silver plastic. No cockpit window openings. No landing gear. Inscribed decal lines. Also issued by Addar in 1976.

298 Douglas B-66 (Co): 1965-68, 1/130; **$20-30**
Ex-Comet. Gray, clear plastic. Inscribed decal lines, no landing gear. A B-66B. Remained in Holland catalog until 1971. Also issued by Frog in 1958 and copied by Marusan. This is the hardest of the ex-Comet kits to find. Box art by Künstler.

299 F-84F Thunderstreak (Co): 1965-68, 1/80; **$15-20**
Ex-Comet. Gray, clear plastic. Only five poorly molded pieces to this kit! Also issued by Frog in 1956 and Addar in 1976. Box art by Künstler.

300 Famous Fighters 4-In-1 Kit 53, 1/48; **$600-700**
Known sets have F9F, F-90, P-40, and ME-109 kits in regular box bottoms, all enclosed in one large box along with a tube of glue. The F9F and F-90 are pictured on the box. The box top has a white square where the contents is rubber stamped so that various kits could be packaged inside. Kits highlighted on the box sides include F-84 Thunderstreak and F-89 Scorpion models that Aurora never issued. $2.98. This very rare kit was issued only during the 1953 Christmas season.

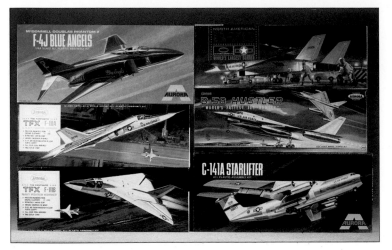

367 F4J Blue Angels: 1970-77, 1/48; $15-25

Reissue of **394** in blue plastic with Blue Angels decals. Also issued as **391**. Box art by Schaare.

368 TFX F-111A 1966-75, 1/48; $30-40

Gray, clear plastic. Includes two crew figures. Has movable swept wings that are geared to move in unison. Retractable landing gear. Air Force version. Also issued as **369**. Box art by Grinnell. Reissued by Monogram.

369 TFX F-111B 1966-69, 1/48; $30-40

Light gray, clear plastic. Navy version of the controversial General Dynamics aircraft the Navy rejected in 1968. Box art depicts one of the seven F-111B's built for carrier trials, but decals are hypothetical. Also issued as **368**. Box art by Grinnell.

370 North American B-70 1962-77, 1/116; $50-70

White, clear plastic. Faint decal locater lines. Includes parts for six GE turbojets, but a very simple kit. Has landing gear, but no gear wells. Early flat box, oval logo issue is more valuable than later square box. Bears the made-up serial number 14005 because the kit was on the store shelves before the first XB-70 was photographed. 1970s issues identified as XB-70. Box art by Leynnwood.

375 B-58 Hustler: 1959-77, 1/76; $30-100

Gray, clear plastic. Decals for Convair's prototype. Raised decal lines. No wheel wells for landing gear. Includes belly weapons pod. First issue box art by Kotula. Second box art by Leynnwood.

376 Ryan X-13 Vertijet: 1959-61, 1/48; $140-160

Silver, clear plastic aircraft. Yellow plastic mobile operational trailer. Raised panel and decal locater lines. Pilot and ground crew figures. Short production life of kit and high demand by collectors make this a valuable kit. Box art by Kotula.

376 C-141 Starlifter: 1969-77, 1/108; $40-60

Silver, clear plastic. Release of this kit was delayed until 1970 because of change in management at Aurora. A simple model with sparse, crude surface detail. Box art by Schaare.

390 Lockheed F-94 Starfire: 1954-58, 1/82; $30-40

Silver, transparent green plastic. Includes landing gear. Heavy riveting and raised decal lines. When issued it was Aurora's lowest priced kit at .39 cents. Catalog number changed to **495** in 1959 when price went to .49 cents.

391 McDonnell Phantom 110: 1965-69, 1/48; $60-70

Gray, clear plastic. The early Air Force version of the Navy's F4U. Also issued as **367, 394**. Box art by Leynnwood.

393 C-119 Flying Boxcar: 1960-77, 1/77; $35-70

Silver, clear plastic. Includes opening cargo doors, loading ramp, jeep, and cannon. First box art by Kotula.

394 F4 Phantom: 1961-69, 1/48; $40-60

Gray, clear plastic. Navy decals. No crew figures. Replaced in catalog with **F4 Blue Angels 367** from 1970-77. Also issued as **F-110 391**. Includes instrument boom in nose. Inaccurate kit based on photos of the prototype, yet a very popular model. Box art by Kotula.

395 A-7D Corsair II 1970-77, 1/48; $25-35

Tan, clear plastic. A Ling-Temco-Vought A-7D. Has retractable landing gear. Includes two Sidewinder missiles. Reissued by Monogram. Box art by Schaare.

490 F-100 Super Sabre: 1954-71, 1/77; $15-35

Silver, transparent green plastic. Includes landing gear. Heavy raised decal lines and rivets.

492 Convair B-36 1957-71, 1/333; $25-35

Silver, clear plastic. Raised decal lines and landing gear. Box art by Kotula.

493 B-47 Stratojet: 1957-71, 1/180; $25-35

Silver, clear plastic. A B-47E. Copied by Marusan of Japan. Box art by Kotula.

493 B-47 Stratojet (Co) 1/205; $25-35

Both original Aurora kit and this ex-Comet kit were boxed under number **493**. Canadian issue **2493**. Box art by Kotula.

494 Boeing B-52 Stratofortress: 1957-71, 1/270; $20-35

Silver, clear plastic. Raised decal lines. A B-52B. Box art by Kotula.

494 B-52 Stratofortress (Co) 1/317; $20-30

Silver, clear plastic. Inscribed decal lines, no landing gear. This ex-Comet kit model is slightly smaller than its Aurora equivalent. Both the original Aurora and ex-Comet models were boxed under number **494**. Canadian issue **2494**. Issued by Frog in 1958. Box art by Kotula.

495 F-94C Starfire: 1959-71, 1/82; **$20-30**
Silver, clear plastic. Raised decal lines, rivets. Issued under **390** from 1954-58. Box art by Kotula.

496 F7U Cutlass: 1962-71, 1/70; **$20-25**
Gray, clear plastic. Big rivets! Box art by Kotula.

2493 B-47 (Co) 1/205; **$20-30**
Ex-Comet kit issued under this number in Canada. Also issued as **493**.

2494 B-52 (Co) 1/317; **$20-30**
Silver, clear plastic. Ex-Comet kit issued under this number in Canada in 1964. Also issued as **494**.

2495 F-94C (Co) 1/92; **$20-30**
Ex-Comet kit issued under this number in Canada. Smaller scale than the original Aurora model (**390**), with recessed decal lines and no rivets. Also issued by Addar in 1976.

Helicopters

350 HC-1B Chinook: 1962-75, 1/48; **$80-100**
Olive, clear plastic. Aurora's first original helicopter kit and its largest. Box art by Kotula.

499 Sikorsky CH-54A Sky Crane: 1969-75, 1/72; **$20-35**
Gray-green, clear plastic. Detachable cargo pod has hinged loading ramp/door and medical equipment inside. Cockpit doors can be assembled in open or closed position. Box art by Schaare.

500 Bell UH-1B Huey: 1966-77, 1/50; **$20-30**
Olive, clear plastic. Includes 4 crew men. Armed with side machine guns and rockets. Box art by Grinnell. Reissued by Allan Brookers of New Zealand in Monogram box in 1980.

501 Hiller Hornet Ram Jet (HI): 1956-65, 1/16; **$130-150**
Olive plastic. No rivets, decal lines, surface panel lines. Windshield is a sheet of flexible, clear acetate. No crew figures. Decals for Air Force, Army, Marines, Navy. Originally issued in 1952 by Helicopters for Industry. This is the world's first helicopter model. Bell AH-1G Cobra also issued under **501**.

501 AH-1G Huey Cobra 'Copter: 1968-77, 1/48; **$20-30**
Olive, clear plastic. Bell helicopter. Hiller Ram Jet also issued under **501**. Box art by Schaare.

502 Piasecki H-25A Army Mule (HI): 1956-65, 1/48; **$90-140**
Olive, clear plastic. Lightly riveted, no decal lines. No crew figures. Originally issued by Helicopters for Industry. Lockheed AH-56A also issued under **502**.

502 Lockheed AH-56A Cheyenne 'Copter: 1969-77, 1/72; **$20-30**
Light tan, clear plastic. One of Aurora's better kits. Nose gun raises and lowers, belly turret rotates. 1968 Flat box issue is rare. Piasecki Army Mule also issued under **502**. Box art by Schaare.

503 Sikorsky S-55 Wind Mill (HI): 1956-69, 1/45; **$80-130**
Dark blue, clear plastic. Pontoon equipped. Lightly riveted, no decal lines. No crew figures. Decals for Air Force, Army, Marines, Navy. Side door slides open. Originally issued by Helicopters for Industry.

504 Piasecki H-21 Work Horse (HI): 1956-69, 1/49; **$80-100**
Silver, clear plastic. Side doors slide open, lift crane swings out. No decal lines, crew figures. Originally issued by Helicopters for Industry. Aurora added rivets and surface detail to the original mold. Second Vietnam-era box art by Steel.

505 Kaman HOK Egg Beater (HI): 1956-65, 1/50; **$160-180**
Dark blue, clear plastic. Heavy rivets, but no decal lines. No crew figures. The overlapping twin rotors are neatly geared to intermesh. Rear fuselage clamshell doors swing open, side windows slide open. Decals for Air Force, Army, Marines, Navy. Originally issued by Helicopters for Industry. Sikorsky HH-3 also issued under **505**.

505 Sikorsky HH-3 Jolly Green Giant: 1969-77, 1/72; **$20-30**
Dark green, clear plastic. *IPMS Journal* (June 1969) said: "As with most Aurora kits, there is very little surface detail, and the rotor has been simplified in the interests of ease of assembly." Has retractable landing gear. Kaman HOK also issued under **505**. Box art by Schaare.

Civilian General Aviation

83 Beechcraft Super 18 (Co): 1965-69, 1/60; **$30-40**
Ex-Comet. White, clear plastic. Although this kit lacks any interior parts or landing gear, it—and the other Comet civilian planes—are cleanly molded with finely inscribed panel lines on the wings and fuselage. Teardrop Comet stand. Box art by Künstler.

85 Aero Jet Commander: 1969-72, 1/72; **$35-45**
White, clear plastic. Nicely detailed interior. Issued as a bagged kit with header card by Aurora, UK in the 1970s. Mold destroyed in train wreck on way to Monogram. Box art by Künstler.

279 Cessna Skymaster: 1969-71, 1/72; **$35-45**
White, clear plastic. Plastic parts have sink marks and faintly inscribed panel lines. Issued as a bagged kit with header card by Aurora, UK in the 1970s. Mold destroyed in train wreck on way to Monogram.

280 Piper Apache (Co): 1964-71, 1/64; **$30-40**
Ex-Comet. Copied by Marusan. Light blue, clear plastic. No landing gear. Issued as a bagged kit with header card by Aurora, UK in the 1970s.

281 Piper Cherokee 180: 1969-71, 1/72; **$30-40**
White, clear plastic. Issued as a bagged kit with header card by Aurora, UK in the 1970s.

282 Piper Aztec C: 1969-71, 1/72; **$40-50**
White, clear plastic. Issued as a bagged kit with header card by Aurora, UK in the 1970s.

283 Cessna 310 (Co): 1964-71, 1/69; **$30-40**
Ex-Comet. Yellow, clear plastic. No landing gear. Issued as a bagged kit with header card by Aurora, UK in the 1970s.

284 Beechcraft Super 18 (Co): 1964-70, 1/88; **$40-50**
Ex-Comet. Silver, clear plastic. No landing gear. Used same box art as larger scale Super 18 **83**. Issued as a bagged kit with header card by Aurora, UK in the 1970s.

285 Aero Commander (Co): 1964-70, 1/81 **40-50**
Ex-Comet. Turquoise, clear plastic. No landing gear, but a finely detailed kit with clear parts for passenger windows. Inscribed decal markings. Issued as a bagged kit with header card by Aurora, UK in the 1970s. Box art by Box art by Künstler.

Airliners

Aurora's airliners are basic models with little surface detail, no clear parts for the passenger window openings, and no wheel wells. After 1973 all of the airliners' passenger window openings were filled-in, and the windows were represented by black squares on a decal strip. Models of aircraft from various airlines were made by issuing the same model packaged with different box art, including appropriate company decals, and sometimes molded in a different color plastic.

127 Russian TU-104 Air Liner: 1959-64, 1/130; **$65-75**
Gray, clear plastic. Aeroflot jet. Nicely molded in a smaller size than most Aurora airliners because Aurora treated it as a "Russian" subject, not an "airliner" kit. Box art by Kotula.

351 Boeing 727 Eastern: 1966-69, 1971-77, 1/96; **$35-45**
Also issued as **353, L-353-H, 354, 355, 357, 382**.

353 Boeing 727 United: 1962-77, 1/96; **$35-45**
Also issued as **351, L-353-H, 354, 355, 357, 382**. Box art by Leynnwood.

353 Boeing 727 Lufthansa: 1974, 1/96; **$45-55**
Issued by Aurora Netherlands. Box art by Leynnwood.

L-353-H Boeing 727 Lufthansa: 1960s 1/96; **$50-60**
Issued by Aurora Netherlands. Also issued as **351, 353, 354, 355, 357, 382**.

354 Boeing 727 TWA 1965-73, 1/96; **$60-70**
Also issued as **351, 353, L-353-H, 355, 357, 382**. Reissued by Monogram.

355 Boeing 727 American 1967-69, 1/96; **$40-50**
Also issued as **351, 353, L-353-H, 354, 357, 382**.

114

356 Douglas DC-9 TWA: 1966-69, 1/72; **$35-45**
Also issued as **357, 386**.

356 Douglas DC-9 Air Canada: 1966, 1/72; **$40-50**
Issued by Aurora Canada in 1966 and again in the 1970s.

356 Douglas DC-9 KLM: late 60s 1/72; **$40-50**
Issued by Aurora Netherlands.

357 Douglas DC-9 Eastern: 1966-70, 1/72; **$35-45**
Also issued as **356, 386**. Box art by Künstler.

357 Boeing 727 Air Canada: late 60s? 1/96; **$40-50**
Issued by Aurora Canada. Also issued as **351, 353, L-353-H, 354, 355, 382**.

358 Boeing 747 Braniff: 1973-77, 1/156; **$35-45**
Orange, clear plastic. Also issued as **360, 361, 362, 363, 379, 383, 391**.

359 Boeing 737 United: 1966-73, 1/72; **$35-45**
Scale Modeler (June 1974) noted poor fit of wings and tail to fuselage and said: "The panel line molding is a little on the heavy side and surface detail is scarce." Also issued as **387**. First issue box art by Grinnell. Second issue box art by Amendola. Reissued by Monogram.

359-1 Boeing 737 Lufthansa: 1974, 1/72; **$55-65**
Issued by Aurora Netherlands.

360 Boeing 747 Delta: 1971-77, 1/156; **$35-45**
Also issued as **358, 361, 362, 363, 379, 383, 391**.

360 Boeing 747 KLM: early 70s 1/156; **$40-60**
Issued by Aurora Netherlands.

361 Boeing 747 Pan Am: 1968-77, 1/156; **$40-50**
Aurora's model was in hobby shops before Pan Am's 747s were in the air. Aurora based decal markings on Pan Am 707s—thus they do not match markings actually adopted by Pan Am. Also issued as **358, 360, 362, 363, 379, 383, 391**. Reissued by Monogram.

361KL Boeing 747 KLM: 1968, 1/156; **$50-60**
Issued by Aurora Netherlands. Ivory, clear plastic.

361-1 Boeing 747 British Air: 1974, 1/156; **$70-80**
Issued by Aurora Netherlands.

362 Boeing 747 United: 1968-77, 1/156; **$40-50**
Also issued as **358, 360, 361, 363, 379, 383, 391**. Box art by Amendola.

363 Boeing 747 TWA: 1971-77, 1/156; **$35-45**
Also issued as **358, 360, 361, 362, 379, 383, 391**. Box art by Künstler.

365 Douglas DC-10 KLM: early 1970s, 1/144; **$45-60**
Issued by Aurora Netherlands. Also issued as **366, 385, 390**.

366 Douglas DC-10 American: 1970-77, 1/144; **$30-40**
Silver, clear plastic. Decals. The DC-10 was the last new aircraft model made by Aurora. Overall quality of this kit is below standard due to tooling cost cutting. Also issued as **365, 385, 390**. Reissued by Monogram. Box art by Schaare.

366-1 Douglas DC-10 SAS: 1974, Netherlands 1/144; **$55-65**

366-2 Douglas DC-10 Martinair: 1974, Netherlands 1/144; **$55-65**
Silver, clear plastic. Decals. Martinair was a Dutch airline.

366-3 Douglas DC-10 Swissair: 1974, Netherlands 1/144; **$55-65**

366-4 Douglas DC-10 Lufthansa: 1974, Netherlands 1/144; **$55-65**

379 Boeing 747 Continental: 1973-77, 1/156 **35-45**
Also issued as **358, 360, 361, 362, 363, 383, 391**.

380 Boeing 707 American: 1966-73, 1/104; **$45-55**
Gray, clear plastic. Decals changed in 70 to match new American markings. Also issued as **381, 382, 395** and in modified form as **720B 388, 396, L-396-H**.

381 Boeing 707 Pan Am: 1959-70, 1/104; **$45-55**
The first of Aurora's airliner kits. Also issued as **380, 382, 395, 388, 396, L-396-H**. Box art by Kotula.

382 Boeing 707 TWA: 1959-69, 1/104; **$35-80**
Also issued as **380, 381, 395, 388, 396, L-396-H**. Box art by Kotula.

382 Boeing 707 Canadian Pacific: early 1970s, 1/104; **$30-40**
Issued by Aurora Canada.

383 Convair 880 Jetliner Delta: 1961-69, 1/103; **$80-110**
White and clear plastic. Also issued as **384**.

383 Boeing 747 Canadian Pacific: early 1970s 1/156; **$35-45**
Issued by Aurora Canada. Also issued as **358, 360, 361, 362, 363, 391**.

384 Convair 880 TWA: 1961-69, 1/103; **$70-100**
Convair built the 880 as its first commercial airliner. Only a few real aircraft and only a few Aurora models were made. Also issued as **383**.

385 Douglas DC-10 United: 1973-77, 1/103; **$25-35**
Also issued as **365, 366, 390**. Box art by Amendola.

386 Douglas DC-8 Jet Clipper Pan Am: 1960-69, 1/103; **$90-100**
America's second jetliner and Aurora's second airliner kit. Also issued as **387, 388, 389, 390**.

386 Douglas DC-9 Hughes Air West: 1974-77, 1/72; **$35-45**
"Sundance" yellow, clear plastic. In 1974 Aurora issued this Air West plane and the PSA and Western airliners in an attempt to boost regional kit sales. Also issued as **356, 357**.

387 Boeing 737 PSA: 1974-77, 1/72; **$35-45**
White, clear plastic. Also issued as **359**.

387 Douglas DC-8 United: 1960-70, 1/103; **$70-90**
Also issued as **386, 388, 389, 390**.

388 Douglas DC-8 Eastern: 1960-69, 1/103; **$80-100**
Also issued as **386, 387, 389, 390**.

388 Boeing 720B Western: 1974-77, 1/104; **$35-45**
White, silver, clear plastic. Having already modified the 707 mold to make the **396** 720-B, Aurora made some additional modifications to make this 720B version more accurate. A fin was added on the bottom of the fuselage, the nose was lengthened, and the passenger windows were filled in and replaced by black squares on the decal sheet to represent windows. Also issued as **396, L-396-H**. Originally issued as Boeing 707 **380, 381, 382, 395**.

389 Douglas DC-8 Delta: 1960-69, 1/103; **$50-100**

389 Douglas DC-8 Garuda: early 60s 1/130; **$300-400**
Also issued as **386, 387, 388, 390**.

390 Douglas DC-10 National: 1973-77, 1/144; **$30-40**.
Decal includes "Barbara" nickname. Box art by Amendola.

390 Douglas DC-8 Air Canada: 66, 1/103; **$60-70**
Issued by Aurora Canada.

390 Douglas DC-8 Trans Canada: 62, 1/144; **$60-70**
Issued by Aurora Canada. Also issued as **386, 387, 388, 389**.

391 Boeing 747 Air Canada: early 70s 1/156; **$40-50**
Issued by Aurora Canada. Also issued as **358, 360, 361, 362, 363, 379, 383**.

395 Boeing 707 Continental: 1962, 1/104; **$100-120**
Also issued as **380, 381, 382, 388, 396, L-396-H**.

396 Boeing 720B Western: 1962-64, 1/104; **$60-80**
Aurora modified its 707 mold slightly to make it a 720B. Engine parts different. Issued with more modifications and new decals as **388**. Also issued as Boeing 707 **380, 381, 382, 388, L-396-H**.

L-396-H Boeing 720B Lufthansa: 1965, 1/104; **$60-80**
Issued by Aurora Netherlands.

397 Convair 990 Jet Mainliner United: 1962-64, 1/107; **$190-220**
White, cream, clear plastic. The kit bears a 1960 copyright, but the instructions and box were not printed until 1962. Convair delayed production and then made only a handful of 990s. Aurora made only a few kits making it the rarest of the domestic airliner models.

Aurora/Heller "Prestige Series"
These kits were produced by Heller in France; packaged and sold in the United States by Aurora.

6600 Super Frelon SA-321: 1977, 1/35; **$20-30**
A large, detailed kit.
6601 F4U Corsair: 1977, 1/72; **$5-6**
6602 Messerschmitt Bf-109G: 1977, 1/72; **$5-6**
6603 P-47N Thunderbolt: 1977, 1/72; **$5-6**
6604 Focke Wulf FW-190F: 1977, 1/72; **$5-6**
6605 Spitfire VB: 1977, 1/72; **$5-6**
6606 Polikarpov I-153: 1977, 1/72; **$5-6**
6607 P-51D Mustang: 1977, 1/72; **$5-6**
6608 Messerschmitt ME-262B: 1977, 1/72; **$5-6**
6609 Messerschmitt Bf-109K: 1977, 1/72; **$5-6**
6610 Yak-3: 1977, 1/72; **$5-6**
6611 Fieseler Fi-156 Storch: 1977, 1/72; **$5-6**
6612 Bucker Bu-133 Jungmeister: 1977, 1/72; **$5-6**

K & B Collectors Series
In 1972 Aurora reissued a set of twelve classic aircraft kits. Manufacturing and sales were handled through K & B, Aurora Canada, and Aurora Netherlands. The molds were improved by removing some of the raised decal lines and in a few other minor ways. The kits were packaged in larger square boxes with vacuum formed bases, rather than the original injection molded ground pieces. New box art was done by John Amendola. Decals and painting instructions were made more accurate. The numbering system added a "1" to the front of each kit's earlier Aurora catalog number.

1100 Sopwith Triplane: 7192, 1/48; **$30-40**
Tan plastic with brown accessory parts. Decal locater markings remain on the wings. Decals for Robert Collishaw's "Black Maria." Box art by Amendola.

1108 Nieuport 28: 1972, 1/48; **$30-40**
Tan, black plastic. The choice of tan plastic was to match the lighter underwing color of the actual aircraft. Hobbyists could leave the under surfaces tan and paint the upper surfaces darker brown or olive. No changes in this mold. Decal for war bonds poster Eddie Rickenbacker painted on his top wing. Box art by Amendola.

1109 Pfalz D-3: 1972, 1/48; **$40-50**
Silver, black plastic. All major decal locater markings removed. Box art by Amendola.

1115 Boeing P-26A: 1972, 1/48; **$30-40**
Yellow, clear plastic. No changes in mold. Decals for 34th Attack Squadron. Box art by Amendola.

1122 Boeing F4B4: 1972, 1/48; **$30-40**
Gray, clear plastic. Seated pilot and standing mechanic. Inscribed decal lines removed. Decals for bomber from USS *Lexington*. Box art by Amendola.

1125 DeHavilland DH-10A: 1972, 1/48; **$40-50**
Tan, black plastic. Only change in this kit was to remove the identification number on the tail and to provide a new number on the decal sheet. Box art by Amendola.

1126 Gotha G.V: 1972, 1/48; **$40-50**
Blue, black plastic. Since Gothas were night bombers, a dark color was appropriate. Decal locater markings removed only from the fuselage. Box art by Amendola.

1134 Fokker E-3: 1972, 1/48; **$30-40**
Tan, brown plastic. No change in the mold. Box art by Amendola.

1135 Fokker E-5: 1972, 1/48; **$35-45**
Gray, black plastic. Decals for the post-war Polish air force. Box art by Amendola.

1136 Halberstadt CL-II: 1972, 1/48; **$40-50**
Red, black plastic. Artist John Amendola said red plastic was chosen to match an unusual, but authentic, paint scheme which would be easier to reproduce than the standard lozenge camouflage. Box art by Amendola.

1141 Breguet 14A: 1972, 1/48; **$50-60**
Tan, black plastic. Bomb racks added, but no bombs. Box art by Amendola.

1142 Albatross C-III: 1972, 1/48; **$55-65**
Tan, black plastic. Decal locater markings removed from fuselage. Box art by Amendola.

Rockets and Missiles

130 Mace TM-76: 1959-60, 1/48; **$80-95**
White plastic. Model has only ten parts. Panel and decal locater lines are faint. It is displayed on a stand. The pattern for a mobile launcher was made but never produced. Box art by Kotula.

131 Bomarc IM-99: 1959-60, 1/48; **$80-95**
White plastic. It is displayed on a stand. Also issued with launcher as **377**. Box art by Kotula.

132 Regulus II: 1959-60, 1/48; **$80-90**
Issued in red plastic, although the real missile was painted blue. Displayed on a stand. Also issued as **378** with launcher. Box art by Kotula.

311 M8E2 Weapons Carrier and Ground Support War Head: 1960, 1/48; **$400-500**
Olive plastic. Aurora's conjectural version of a LaCrosse missile added to the M8E2 Munitions Carrier **309** with plow blade, ammunition lift, and other details removed. Guard rails, ladders, and pneumatic launcher tube added. Launcher raises and lowers. Driver and observer figure.

377 Bomarc IM-99: 1959-60, 1/48; **$280-300**
White plastic missile, operating silver plastic mobile launcher. Chassis of launcher is same as that of **378 Regulus**. Since actual missile was black, the kit includes decals for either a white or black color scheme. Issued without launcher as **131**.

378 Regulus II: 1959-60, 1/48; **$190-210**
Red plastic missile, operating gray and silver plastic mobile launcher. Chassis of launcher is same as that of **377 Bomarc**. Decals. Three ground crew figures. Actual Navy missile was blue. Issued without launcher as **132**. Box art by Kotula.

379 Nike Hercules: 1959-60, 1/48; **$180-200**.
White plastic rocket, operating olive launching platform. Four ground crew figures. Box art by Kotula.

380 IRBM Thor: 1959-60, 1/72; **$280-300**
White rocket, gray plastic launch platform and service tower. Three ground crew figures. Decals. Insulated wire for fuel and oxygen lines. Launch platform is the same as that in **385 Lunar Probe**.

385 USAF Lunar Probe: 1959-60, 1/72; **$280-300**
White rocket, gray plastic launch platform and service tower. Insulated wire for fuel and oxygen lines. Decals. Launch platform is the same as that in **380 Thor**. The US created the Thor-Able rocket by adding three stages to the Thor; Aurora added extra parts to its **380 Thor** model. Nose cone lifts off to show the "Pioneer" probe.

Popsicle/Aurora Aircraft and Rocket Trading Cards

The cards were individually packaged with Popsicle's frozen treats and bear the same art work as Aurora's model kit boxes. Aurora and Popsicle announced release of the cards in September, 1959. They came in two sets of twenty-one cards each. Cards in the first set have the Aurora logo and information on the Aurora kit on the back. The second set does not. These cards are very hard to find and sell for about **$6** each. Cards are listed by their card number, with the Aurora kit number in parentheses.

F86D SABRE JET

A- 1 F8U-1 Crusader (119)
A- 2 B-47 Stratojet (493)
A- 3 F9F-6 Cougar (293)
A- 4 B-52 Stratofortress (494)
A- 5 Convair F-102 (290)
A- 6 F-104 Starfighter (291)
A- 7 F-100 Super Sabre (289)
A- 8 B-36 Convair (492)
A- 9 F-90 Lockheed (33)
A-10 F9F Panther Jet (22)
A-11 F-94C Starfire (390)
A-12 F-86D Sabre Jet (77)
A-13 X-13 Vertijet (376)
A-14 F-101 Voodoo (294)
A-15 North American F-107A (295)
A-16 F-105 Thunderchief (123)
A-17 X-15 Sateloid Plane (120)
A-18 B-58 Hustler (375)
A-19 Bomarc IM-99 (131)
A-20 Regulus II (132)
A-21 Nike-Hercules (379)
A-22 F4D Skyray (292)
A-23 Convair Pogo (60)
A-24 Lockheed VTO (50)
A-25 Martin B-26 (271)
A-26 CF-105 (124)
A-27 Boeing B-29 (372)
A-28 North American B-25 (373)
A-29 Piasecki H-25A (502)
A-30 Kaman HOK (505)
A-31 Hiller Hornet (501)
A-32 Sikorsky S-55 (503)
A-33 Piasecki H-21 (504)
A-34 Mace TM-76 (130)
A-35 Thor IRBM (380)
A-36 Lunar Probe (385)
A-37 P-38 (99)
A-38 P-51 Mustang (118)

A-39 Curtiss P-40 (44)
A-40 Grumman Hellcat (40)
A-41 B-17 Flying Fortress (491)
A-42 Spitfire (20)

Ships

Historic Sailing Ships

210 Black Falcon: 1954-75, 1/100; **$25-65**
Early issues are ivory plastic with metal anchor chains; later issues are black plastic. A toylike model of a pirate ship based on a wooden model by Ideal Aeroplane Supply of New York City. One of Aurora's best selling kits. Also issued as **429**. First two box art by Cox. Third rowboat box art by unknown artist. Final box art by Steel. Copied by Merit of England in the late 1950s, this pirated mold was later used by Artiplast of Italy and SMER of Czechoslovakia.

320 Viking Ship: 1954-75, 1/80; **$25-65**
Early issues ivory, later brown plastic, vinyl sail. First issue has eagle sail decal, later a dragon. Includes numerous crew figures. Based on a wooden model by Ideal. Copied by Merit, also reissued by Artiplast and SMER. First box art by Cox. Second box art with two ships by unknown artist. Final box art by Steel.

429 Buccaneer Sea Raider: 1965-70, 1/100; **$65-75**
Dark brown plastic, metal anchor chains. Reissue of **210** with new box reading, "Inspired by the Great Motion Picture, Cecil B. DeMille presents *The Buccaneer*." This kit was an attempt to capitalize on the 1958 remake of *The Buccaneer* starring Yul Brynner and Charlton Heston. Box art done by Jo Kotula and box is dated 1959. The kit appears on a 1960 dealers order sheet, but does not appear in the catalog until 1965.

430 Chinese Junk: 1955-70, 1/68; **$25-75**
Black plastic, orange injection molded sails. Includes decal for stern art and paper pendants. Reissued as **437**. First box art by Cox. Second box art by Steel.

431 Schooner Bluenose: 1958-73, 1/124; **$20-40**
Green, tan plastic. Vacuum formed sails, metal anchor chain. Paper Canadian flag. Classic Canadian fishing boat that won races off New England in the 1930s. First box art by Kotula. Second box art by Steel.

 432 Cutty Sark: 1959-75, 1/260 **Corsair**; **$20-40**
Dark green, tan, black plastic. Paper flags. No sails. Mid-19th century clipper ship. First box art by Kotula. Second box art by Steel.

433 Privateer Corsair: 1959-70, 1/128 **Corsair**; **$50-70**
Brown, black plastic, vacuum formed sails. Armed sloop from Revolutionary War era. First box art by Kotula. Second box art by Steel.

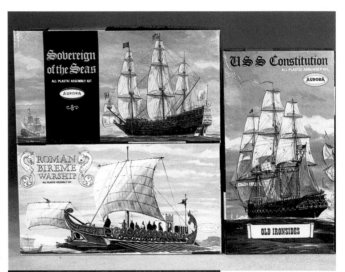

434 Sovereign of the Seas: 1967-73, 1/260; **$30-40**
Brown plastic, injection molded sails. Paper flag sheet. 100 gun British warship launched in 1637. Box art by Steel.

435 Roman Bireme: 1967-70, 1/80; **$25-35**
Light brown plastic, injection molded sail. The Academy/Minicraft "Roman Warship" is a larger scale model of the same ship. Heller's "Bireme Imperator" is another model of the same craft in 1/75 scale, and IMAI made yet a fourth model of this same ship in small scale. Box art by Steel.

436 USS Constitution: 1967-73, 1/260; **$20-30**
Brown plastic, black ratlines, injection molded sails. Paper flag sheet. Box art by Steel.

437 Armed Command Junk: 1968-69, 1/68; **$65-75**
Reissue of Chinese Junk **430**. Black and orange plastic with new tan plastic parts to convert model into a Vietnam War ship. Machine guns are concealed under bales and tarpaulins that lift off. Box art by Steel.

440 Wanderer Whaler: 1966-73, 1/120; **$50-70**
Brown, tan, black, white plastic. First of the four large sailing ships issued in 1966. Box art by Steel.

441 USS Hartford: 1966-73, 1/115; **$60-80**
Gray plastic hull, brown deck, black detail parts, white injection molded sails, metal anchor chains, string for rigging. Admiral Farragut's flagship from the Civil War. Box art by Steel.

442 Sea Witch: 1967-73, 1/118; **$60-80**
Dark brown, light brown, black, white plastic. Metal chains, string, paper flag sheet. Clipper ship. Box art by Steel.

443 Bon Homme Richard: 1967-73, 1/118; **$60-80**
Dark brown, light brown, black plastic. Injection molded sails, metal anchor chains, string for rigging. Decals for captain's cabin windows. John Paul Jones's ship. Box art by Steel.

202 Naval Battles of World War II: 62, 1/600; **$350-370**
Released in cooperation with *American Heritage* magazine. Contains **713 Yamato** and **714 Enterprise**. Box says it also contains a booklet on sea battles of World War II, but no existing kits have this booklet.

444 Wheeler Cruiser: 1968-70, 1/48; **$70-80**
White, brown, blue, and chrome plastic. Five figures. A 65 foot sport fishing boat. 17 inch long model. An expensive kit at $5. Box art by Künstler.

480 USS Halford: 1954-64, 1/305; **$70-90**
Gray plastic. Decals, paper signal flags sheet. Aurora's kit 480 is a model of DD 480 *Halford*. Or is it? The real *Halford* was a Fletcher class destroyer, with an experimental scout plane catapult in place of one of its aft gun turrets. Aurora's model is a Gearing class destroyer, with an aircraft catapult that no Gearing carried. First box art by Cox. Second box art by Kotula.

500 Nautilus: 1953-56, 1/242; **$150-160**
Dark gray plastic. Includes small tube of cement. Aurora's first ship model. First edition has Brooklyn address and reads "U-Ma-Kit" on side panel. Based on drawings in an article in the December 1952 issue of *Colliers* magazine. A very simple kit. Includes decal sheet with shark's mouth and eyes from P-40 aircraft model. Reissued as **708** in 1959. Also reissued as **615**. Box art by Cox.

HM698 History in Miniature: 1955-60; **$400-500**
Gift set contains **210, 320, 430**. Includes peel and stick labels for box top reading: "Birthday, Get Well, Fathers Day, etc." Captain Kidd's Pirate Ship, Leif Ericson's Viking Ship, Genghis Khan's Chinese Junk.

614 Skipjack: 1971-72, 1/228; **$70-80**
Gray plastic. Reissue of **711** with "Phota-Scope" magnifying glass peep viewer of photographic slide showing interior of sub. Viewer placed under hatch formerly covering the nuclear reactor. The transparency is a photo taken by Jim Keeler of the German World War II U-505 at the Chicago Museum of Science and Technology! Because the Navy refused to furnish modern submarine interior photos, Keeler took the U-boat tour and snapped photos as he went! Box art by Steel.

615 Nautilus: 1971-72, 1/242; **$70-80**
Black plastic. Reissue of **708** with "Phota-Scope" peep viewer of interior of sub. Viewer placed through the sides of the hull! Another U-505 interior photo. Box art by Steel.

703 St. Paul: 1957-77, 1/600; **$25-70**
Gray plastic. Baltimore class heavy cruiser from World War II. Decal numbers for all Baltimore class ships. First issue box art by Kotula. Second box art by Steel.

704 Bennion: 1957-68, 1/516; **$25-55**
Gray plastic. A Fletcher Class destroyer from World War II. Kit contains names and decal numbers for 55 different ships in this class. Kit released along with Aurora's other 1/600 World War II ships, yet it is in larger scale—probably because a 1/600 destroyer would be too small for a satisfactory model. First box art by Kotula. Second box art by Steel.

705 Iowa: 1957-77, 1/600; **$20-70**
Gray plastic. Decals for USS *Iowa, Missouri, New Jersey.* Iowa class battleship. Original 1957 Kotula box art shows helicopters on aft deck, but kit actually has World War II scout planes. Also issued under **723, 724.** First box art by Kotula. Second box art by Steel.

700 Independence: 1960-77, 1/600; **$25-45**
Gray plastic. Aircraft in white; deleted from later issues. Same mold as **701, 702** with different decals and box. First box art by Ed Marinelli. Second box art by Steel.

701 Forrestal: 1957-77, 1/600; **$25-70**
Gray plastic. Early issue aircraft in blue, later issues in white plastic, last issues delete the now-obsolete aircraft. Decal sheet for Saratoga and Forrestal. Same mold as **700, 702.** First box art by Kotula. Second issue box art by Steel.

702 Saratoga: 1957-77, 1/600; **$25-70**
Gray plastic. Aircraft in white, deleted in latter issues. Same mold as **700, 701.** Contains decal sheet with markings for Forrestal and Saratoga. First box art by Ed Marinelli. Second box art by Steel.

121

706 Sea Wolf: 1957-72, 1/268; **$30-50**
Black plastic. Contains deck hanger and generic missile. Decal sheet has markings for *Sea Wolf* and *Nautilus*. The Navy tested air breathing missiles on submarines, but not on the *Sea Wolf*. The model is identical to the *708 Nautilus*, but was manufactured from a duplicate mold. The real *Sea Wolf* looked quite different from the *Nautilus*. Its first commander, James E. Carter, later became President of the United States. First box art by Kotula.

708 Nautilus: 1959-77, 1/242; **$30-80**
Black plastic. Also issued as **500, 615**. Includes deck hanger and missile. Decals sheet includes *Sea Wolf* markings. First box art by Kotula. Second box art by Steel.

709 Graf Spee: 1959-77, 1/600; **$30-65**
Gray plastic. German pocket battleship sunk early in World War II. First box art by Kotula. Second box art by Steel.

710 Atlantis: 1959-64, 1/456; **$150-200**
Gray plastic. German freighter, converted to a commerce raider during World War II. Includes a scout plane hidden in the forward hold, and cannon hidden under "dummy" deck cargo or behind fold-down hull doors. False smokestack extensions can be added to disguise the ship's profile. Reissued in England and Holland in 1973. First issue box art by Kotula. Second issue box art by Steel.

711 Skipjack: 1960-77, 1/228; **$25-60**
Black plastic. Decals. Raised "585" decal locater markings. Hatch (not on the real sub) opens to reveal nuclear power plant. Propeller, periscope, rudders, and diving planes move. Ken Hart (*Scale Ship Modeler* November 1994) says: "Given the cloak of secrecy surrounding *Skipjack* the kit was not quite accurate, but was surprisingly close." Major errors: three-bladed rather than correct five-bladed prop and too wide diving planes. First box art by Kotula. Second box art by Steel. Reissued as **614** and by Monogram.

712 King George V: 1960-69, 1/600; **$80-100**
Gray plastic. British battleship that tracked down the *Bismarck*. Released in same year as movie *Sink the Bismarck*. Reissued in Holland in 1974. Box art by Kotula.

713 Yamato: 1960-70, 1/550; **$40-60**
Gray plastic. Paper flag sheet. Japanese superbattleship sunk late in World War II. Included in **202**. Reissued in Holland in 1974. Box art by Kotula.

714 Enterprise: 1960-77, 1/600; **$50-150**
Gray plastic. Decals. Paper American flag. Includes only three aircraft. Included in **202**. World War II aircraft carrier not to be confused with the later nuclear carrier model **720**. Box art by Kotula.

715 Bismarck: 1962-77, 1/600; **$35-60**
Gray plastic. World War II German battleship. Box art by Steel.

716 Wolfpack U-boat: 1962-77, 1/209; **$20-40**
Black plastic. Based on U-505 captured in Atlantic in June 1944 and put on exhibit at Chicago's Museum of Science and Industry in 1954. Reissued as **725**. Box art by Steel. Reissued by Monogram.

717 USS Bainbridge: 1965-77, 1/600; **$25-35**
Gray plastic. Nuclear powered frigate with twin Terrier missile launchers fore and aft.

718 USS Guadalcanal: 1967-77, 1/600; **$30-40**
Gray plastic. Six Sea Knight and six Sea King helicopters. Includes complete interior hanger deck which can be seen through sliding doors. Elevators can be assembled in an up or down position.

720 USS Enterprise: 1961-77, 1/391; **$125-150**
Gray, light gray, bronze, chrome. Metal drive shafts for props. Includes 40 aircraft: Crusaders, Phantoms, Skyhawks, helicopters. Nuclear powered supercarrier. Features movable elevators, pre-painted hull and flight deck, chrome plated stand. 33" long. Reissued as **721** and by Monogram. Came in motorized version in 61 at $11.95 and non-motorized at $9.95. A **Motorizing Kit** (**7-198**) was also sold separately. Motors came from Johnson Electric of Hong Kong, which Aurora later bought to manufacture motors for slot cars.

721 USS Enterprise CVAN-65: 1976-77, 1/391; **$90-110**
Gray, light gray, dark gray, black plastic. Revised issue of **720**. Revisions: No pre-painted parts. Flight deck molded in dark gray. Stand in black plastic. New decal sheet.

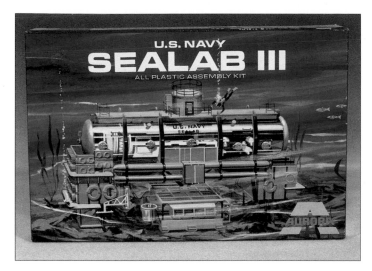

721 Sealab III: 1970-71, 1/93; **$300-350**
Molded in orange plastic with alternative clear hull half to show interior. Light green plastic base. Two scuba diver figures. A $4 kit. Ocean floor base reissued with **Seaview 253**.

726 Russian Guided Missile Submarine: 1968-77, 1/200; **$30-50**
Gray plastic. "Golf I" diesel sub. Simple kit, based on aerial photographs of the sub on the surface; the below waterline features are conjectural.

727 Tucumcari Hydrofoil: 1970-72, 1/84; **$95-105**
Dark gray, light gray plastic. Prototype of the Boeing water-jet powered gunboat PGH-2. The foil boom can be raised and lowered.

728 Japanese Submarine I-19: 1970-77, 1/275; **$25-35**
Dark gray plastic. Paper flag sheet. A B1 class submarine with a deck hanger and "Glen" scout aircraft, which bombed Oregon in 1942. The I-19 sank the US aircraft carrier *Wasp*, and then was sunk by a US destroyer. *Scale Modeler* (September 1970) wrote: "Without a doubt Aurora has come up with the finest submarine kit ever introduced to the market." Reissued by Monogram. Second box art by Schaare.

722 Moscow: 1971-77, 1/600; **$25-35**
Gray plastic. Eight Kamov KA-25 helicopters. Russian helicopter carrier. Box art by Amendola.

723 USS Missouri: 1973-77, 1/600; **$20-30**
Gray plastic. Also issued as **705, 724**. Box art by Steel.

724 USS New Jersey: 1973-77, 1/600; **$20-30**
Gray plastic. Also issued as **705, 723**. Box art by Steel.

725 U-156 German U-Boat: 1973-77, 1/209; **$20-30**
Black plastic. Reissue of **716**. Reissued by Monogram.

598 U. S. Navy Task Force: 1959-60; **$400-500**
Gift set includes **701, 703, 704, 705**.

Aurora/Heller Ships

These kits were manufactured by Heller in France; packaged in the United States by Aurora.

6501 Corona: 1977, 1/200; **$20-25**
6502 Stella del Norte: 1977, 1/200; **$20-25**
6503 Santa Maria: 1977, 1/90; **$20-25**
6504 Drakkar Oseberg: 1977, 1/60; **$15-20**
6515 Bireme Imperator: 1977, 1/75; **$25-35**
6521 Phenix: 1977, 1/200; **$20-25**
6522 Royal Lewis: 1977, 1/200; **$20-25**
6523 Sirene: 1977, 1/200; **$20-30**
6541 La Reale France: 1977, 1/75; **$20-30**
6542 Chebec: 1977, 1/50; **$20-30**
6550 Soliel Royal: 1977, 1/200; **$50-60**

Military Armor and Vehicles

All Aurora's armor models are in 1/48 scale. They all have flexible vinyl treads or wheels that turn, cannon that elevate, turrets that rotate, and hatches that open. Each kit includes at least one crewmen figure and most have escorting infantrymen.

First Series: 1956-69

Flat boxes, rectangular and oval logos. Black vinyl treads.

206 3 Army Tanks: 1958-59, 1/48; **$400-500**
Gift Set includes **301, 302, 303**.

300 Centurian: 1960-69, 1/48; **$25-30**
Olive plastic. Decals. World War II British heavy tank. Commander with walkie-talkie stands in hatch, kneeling infantryman. No decals. Reissued as **330, 061**. First box art by Kotula. (Kotula used himself as the model for the infantryman in the foreground.) Second box art by Steel.

301 M-46 Patton: 1956-69, 1/48; **$25-50**
Olive plastic. Korean War tank. Includes two drivers, observer with binoculars, commander holding microphone. Machine gun located to rear of hatch on turret. Reissued as **321, 062**. Second box art by Steel.

302 Panther: 1956-69, 1/48; **$25-50**
Gray plastic. Tank commander in turret hatch and two drivers. Reissued as **322, 071**. Reissued by Monogram in **6035 Tank Hunter**. Second box art by Steel.

303 Stalin: 1956-69, 1/48; **$25-50**
Green plastic. Machine gunner figure in hatch. Reissued as **323, 070, 612**. Second box art by Steel.

307 8 inch Howitzer: 1959-60, 1964-68, 1/48; **$30-60**
Olive plastic. Four gun crew figures. Except for the cannon barrel itself, this is the same kit as **308**. First box art by Kotula. Second box art by Steel.

308 155 MM Long Tom: 1959-60, 1964-68, 1/48; **$30-60**
Olive plastic. Four gun crew figures. Except for the cannon barrel itself, this is the same kit as **307**. Reissued as **331**. First issue box art by Kotula. Second issue box art by Steel.

309 M8 Munitions Carrier: 1959-60, 1964-69, 1/48; **$30-50**
Olive plastic. Driver, gunner with binoculars, seated infantryman. Ammunition loader at rear raises and lowers, ramp to main cargo bay lowers, scraper blade in front raises and lowers. Reissued as **332, 065**. First issue box art by Kotula. Second issue box art by Steel.

310 8 inch Howitzer With M8 Munitions Carrier: 1959-60, 1/48; **$100-120**
Olive plastic. Reissued as **333**. Box art by Kotula.

312 Tiger Tank: 1964-69, 1/48; **$25-30**
Gray plastic. Tank commander stands in hatch looking through binoculars. Three infantrymen. Reissued as **324, 064, 611**. Box art by Steel.

313 Japanese Medium Tank: 1964-69, 1/48; **$25-30**
Olive plastic. Tank commander in hatch with uplifted arm and three infantrymen. Reissued as **325, 073**. Box art by Steel.

314 M109 Howitzer Tank: 1965-69, 1/48; **$25-30**
Olive plastic. Self-propelled howitzer from 1960s. Commander in hatch, driver, infantryman with walkie-talkie. Reissued as **326, 063**. Box art by Steel.

315 British Churchill: 1967-69, 1/48; **$25-30**
Olive plastic. Black vinyl radio antennas. Tank commander in hatch looking through binoculars. Three infantrymen. Reissued as **327**. Box art by Künstler.

316 Swedish S Tank: 1967-69, 1/48; **$25-30**
Olive plastic. 1960s vintage Bofars main battle tank. Tank commander in hatch and three infantrymen. Reissued as **328**. Box art by Künstler.

317 Sherman: 1968-69, 1/48; **$25-30**
Dark olive. Tank commander in hatch and three infantrymen. *Scale Modeler* (February 1969) said: "In recent years, this company has steadily gone downhill in regards to the quality of their aircraft models, yet their tanks remain excellent; witness the new 1/48 scale 'easy eight' Sherman." Box art by Steel. Reissued as **329, 072**. Reissued by Monogram in **6034 Ground Attack**.

Second Series 1970-75

White square boxes with large "A" logo. Kits include vacuum formed "battlefield display stands." Black vinyl treads. New decal sheets include nameplate for base. Eight kits in this series received new box art. In the other six, John Steel's art was retained, but modified to fit the square format.

318 MBT-70 Main Battle Tank: 1969-75, 1/48; **$30-35**
Dark olive plastic. Four-pronged plastic leaf spring allows hull to move, simulating suspension system of the real tank. Tank commander in hatch and three infantrymen. Alternate decals for US and West German army. The joint German-US MBT-70 project was cancelled shortly after Aurora brought out its kit. No other company has made a model of this tank. Box art by Schaare. Reissued as **060**.

321 General Patton Tank: 1970-75, 1/48; **$25-30**
Olive plastic. Retouched Steel art. Also issued as **301, 062**.

322 German Panther Tank: 1970-75, 1/48; **$25-30**
Tan plastic. Decal markings for three Panzer units. Retouched Steel art. Also issued as **302, 071** and by Monogram in **6035**.

323 Russian Stalin Tank: 1970-75, 1/48; **$25-30**
Olive plastic. Box art by Schaare. Also issued as; **$303, 070, 612**.

324 German Tiger Tank: 1970-75, 1/48; **$25-30**
Light brown plastic. Decal markings for three Panzer units. Retouched Steel art. Also issued as **312, 064, 611**.

325 Japanese Medium Tank: 1970-75, 1/48; **$25-30**
Olive plastic. Box art by Schaare. Also issued as **313, 073**.

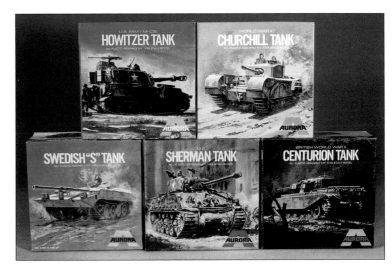

326 M-109 Howitzer Tank: 1970-75, 1/48; **$25-30**
Dark olive plastic. Decals for US and West German armies. Retouched Steel art. Also issued as **314, 063**.

327 Churchill Tank: 1970-75, 1/48; **$25-30**
Olive plastic. Decals for three World War II British units. Box art by Schaare. Reissue of **315**.

328 Swedish "S" Tank: 1970-75, 1/48; **$25-30**
Dark gray plastic. Box art by Schaare. Reissue of **316**.

329 Sherman 1970-75, 1/48; **$25-30**
Olive plastic. Decals for two different US divisions and one Israeli unit. *Scale Modeler* (October 1969) pointed out that Aurora's Sherman tank sold for $1.30, compared to Revell's 1/40 scale Sherman at $2.00 and MRC/Tamiya's 1/35 scale kit at $4.00. Retouched Steel art. Also issued as **317, 072** and by Monogram in **6034**.

330 British Centurian: 1970-75, 1/48; **$25-30**
Olive plastic. Decals for World War II, Korean War, and Israeli army. Retouched Steel art. Also issued as **300, 061**.

Diorama Kits

339 Anzio Beach: 1968-73, 1/87; **$70-80**
Toy soldiers set with injection molded battlefield scenery, LST, German AA Tank, German Panther, US Sherman, US Patton, Jeep. Vehicles are from Roco molds. Germans in gray, Americans in olive, and scenery in tan. Box art by Steel.

340 Rat Patrol: 1968-72, 1/87; **$150-175**
Toy soldiers set with injection molded battlefield scenery. Olive and tan plastic. Vehicles are from Roco molds. Two jeeps, Panzer tank, Panther tank, 15 figures. Oasis with palm trees and sand dunes. Based on TV series *Rat Patrol*. Appears in 1973 catalog as "Desert Raiders," but not issued under this name. These tanks were also sold separately on blistercards as Military Midgets. First box art Harry Schaare painted for Aurora.

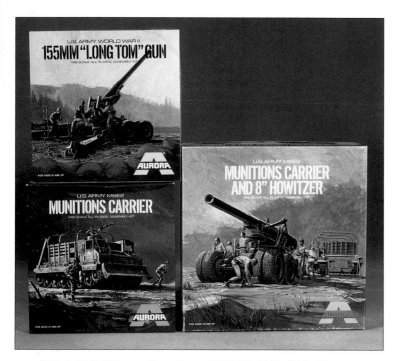

331 155 MM Long Tom Gun: 1973-75, 1/48; **$30-40**
Olive plastic. New art. Reissue of **308**.

332 M8E2 Munitions Carrier: 1973-75, 1/48; **$25-30**
Olive plastic. New art. No vacuum formed base. Also issued as **309, 065**.

333 M8E3 Munitions Carrier and 8" Howitzer: 1973-75, 1/48; **$40-45**
Olive plastic. Larger box than the rest of this series. New art. No vacuum formed base. Reissue of **310**.

Phota-Scope Kits

611 German Tiger Tank: 1971-72, 1/48; **$40-50**
Brown plastic. Phota-Scope kit. Includes vacuum form base. Peep viewer with magnifying glass lens and photographic transparency. Open the turret hatch and look in to see a picture of an explosion. Also issued as **312, 324, 064**.

612 Russian Stalin Tank: 1971-72, 1/48; **$40-50**
Pale green plastic, brown vacuum form base. Phota-Scope kit. Includes same explosion scene as Tiger Tank. Also issued as **303, 323, 070**.

Third Series 1976-77

Flat white box with block "AURORA" logo. Plastic in most kits given surface texture to simulate metal; some kits received minor mold changes. Vinyl tread color changed from black to silver. Kits not reissued in this series are Churchill, Swedish S, 8 inch Howitzer, 155 MM Long Tom.

060 MBT-70: 1976-77, 1/48; **$25-30**
Dark olive plastic. Antenna added to turret, detail and texture added to hull. Decals for US and West German armies. Also issued as **318**.

061 Israeli Centurian V: 1976-77, 1/48; **$25-30**
Tan plastic. No texture added to this kit's surfaces. New 105 mm cannon replaces earlier 85 mm cannon. Rear deck details changed, second radio antenna added to turret, new machine gun in main cannon mantlet. Decals for 1967 war. Also issued as **300, 330**.

062 M-46 Patton: 1976-77, 1/48; **$15-20**
Olive plastic. Machine gun moved from back to front of turret top, and two radio antennae added to rear of turret. Outer rows of exhaust vents on top of rear hull closed over. Surface texture added. Also issued as **301, 321**.

063 M-109 S. P. Howitzer: 1976-77, 1/48; **$15-20**
Dark olive plastic. Decals for US and West German armies. Surface texture added. Also issued as **314, 326**.

064 King Tiger: 1977, 1/48; **$15-20**
Dark gray plastic. Overall surface not textured, but rough areas representing "Zimmerit" anti-magnetic mine paste added to hull and turret sides. Same decals as second issue. Also issued as **312, 324**.

065 M8E2 Munitions Carrier: 1977, 1/48; **$15-20**
Olive plastic. No mold changes except adding of texture. Also issued as **309, 332**.

070 Soviet JS III Stalin: 1976-77, 1/48; **$15-20**
Brown plastic. Surface texture added and machine gun modified from earlier issues. Also issued as **303, 323, 612**.

071 Pzkpfw. V Panther A: 1976-77, 1/48; **$15-20**
Gray plastic. Overall surface not textured, but rough areas representing "Zimmerit" anti-magnetic mine paste added to hull and turret sides. New cannon, towing cables, and antenna on rear of hull. Also issued as **302, 322, 611** and by Monogram.

072 M4A3E8 Sherman: 1977, 1/48; **$15-20**
Dark brown plastic. Decals for US and Israeli armies. Surface texture added. Also issued as **317, 329** and by Monogram.

073 Japanese Type 97 Chi-Ha: 1977, 1/48; **$20-25**
Dark green plastic. Surface texture added. Also issued as **313, 325**.

Aurora/ESCI Military Vehicles and Armor

Manufactured in Italy by ESCI; packaged in United States by Aurora.

6201 King Tiger: 1977, 1/72; **$5-6**
6202 Panzer IIIM/N: 1977, 1/72; **$5-6**
6203 Panzer IVH: 7197, 1/72; **$5-6**
6204 Panther G: 1977, 1/72; **$5-6**
6205 M4A1 Sherman: 1977, 1/72; **$5-6**
6206 M12 SPG: 1977, 1/72; **$5-6**
6207 M3A1 Scout Car: 1977, 1/72; **$5-6**
6208 T-34/76 C-D: 7197, 1/72; **$5-6**
6209 Sdkfz 250/9 Recon Half-Track: 1977, 1/72; **$5-6**
6212 Nebelwerfer Battery: 1977, 1/72; **$5-6**
6213 Sturmgeschutz IIIG: 1977, 1/72; **$5-6**
6214 Elefant: 1977, 1/72; **$5-6**
6215 Opel Blitz: 1977, 1/72; **$5-6**
6216 Jagdpanther: 1977, 1/72; **$5-6**
6217 Sdkfz II Munitions Carrier: 1977, 1/72; **$5-6**
6218 German Weapons Set: 1977, 1/72; **$5-6**
6219 Fiat-Ansaldo M.13/40: 1977, 1/72; **$5-6**
6220 3/4 Ton Weapons Carrier: 1977, 1/72; **$5-6**
6221 Bishop SPG: 1977, 1/72; **$5-6**
6222 Panzerjager I: 7197, 1/72; **$5-6**
6223 Valentine III: 7197, 1/72; **$5-6**
6225 Hanomag Rocket Launcher: 1977, 1/72; **$5-6**
6226 Sdkfz 250/3 Panzer IB: 1977, 1/72; **$5-6**
6227 Sdkfz 250/3: 1977, 1/72; **$5-6**
6228 KV-10 Type C: 1977, 1/72; **$5-6**
6229 M6 Anti-Tank Gun: 1977, 1/72; **$5-6**
6351 Harley-Davidson WLA-45: 1977, 1/9; **$15-20**
6352 Triumph 3 HW: 1977, 1/9; **$15-20**
6353 Zundapp KS-750/1: 1977, 1/9; **$15-20**
6354 BMW R-75/A-1: 1977, 1/9; **$15-20**
6361 BMW R-75 with sidecar: 1977, 1/9; **$15-20**
6362 Zundapp KS-750 with sidecar: 1977, 1/9; **$15-20**
7002 Harley-Davidson 45: 1977, 1/9; **$15-20**
7003 Zundapp 750 with sidecar: 1977, 1/9; **$15-20**
7004 British Triumph: 1977, 1/9; **$15-20**
7005 Kittenrad: 1977, 1/9; **$25-30**
7006 Zundapp 750: 1977, 1/9; **$15-20**
7008 BMW R-75/A1: 1977, 1/9; **$15-20**
7009 Kubelwagen: 1977, 1/9; **$25-30**

Automobiles and Trucks

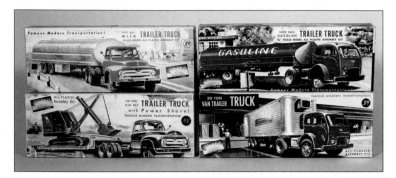

1/64 Trucks

These simple truck kits were declared to be 1/48 scale, but are actually about 1/64. The box side reads "Use in Popular Train Layouts." They have black vinyl tires, but no clear or chrome parts. All boxes have rectangular "Aurora Line" logos.

204 4 Model Trucks: 1958, 1/64
This cellophane-wrapped set was advertised only in 1958. The ad shows it to contain all four of the trucks. No copies of this set are known to exist.

681 5000 Gal. Milk Truck: 1956-57, 1/64; $100-120
Blue plastic. Ford truck with hood that opens to show engine. Silver tank trailer. Decal sheet for "Dairy Milk" and "Gasoline."

682 5000 Gal. Gasoline Trailer Truck: 1956-57, 1/64; $100-120
Red plastic. Cab-over-engine White truck with tank trailer. Cab pivots forward to show engine. "Gasoline" decal. Reissued as 305.

683 40 Ton Trailer With Power Shovel: 1956-57, 1/64; $125-150
Red plastic Ford truck with hood that opens to show engine. Yellow plastic trailer, and orange plastic "Bay City" power shovel. Decals for "Ace Construction Co." Reissued as 306.

684 20 Ton Van Trailer: 1956-57, 1/64; $100-120
Red plastic cab-over-engine White truck. Silver trailer with "Aurora Coast to Coast" decal. Cab pivots forward to show engine. Reissued as 304.

304 U. S. Army Depot Supply Trailer: 1958-59, 1/64; $110-130
Reissue of civilian 20 Ton Van Trailer 684 in army olive plastic. Army decals.

305 U. S. Army 5000 Gal. Gas Truck: 1958-59, 1/64; $110-130
Reissue of civilian 5000 Gal. Gasoline Truck 682 in army olive plastic. One standing, two seated figures added. Army decals.

306 U. S. Army Engineers Power Shovel and Transporter: 1958-59, 1/64; $140-160
Reissue of civilian 40 Ton Trailer With Power Shovel 683 in army olive plastic. Two crew figures added. Army decals.

Ex-Best Cars
All have rubber tires and decals, but no clear or chrome parts.

521 Monroe Special (Be): 1959-63, 1/30; $40-50
Ex-Best. Blue plastic. Includes seated driver and mechanic figures. "4" decals. Tires read "Oldfield Corp." 1920 winner, built by Louis Chevrolet. Box art by Anthony Rudisill.

522 Murphy Special (Be): 1959-63, 1/30; $40-50
Ex-Best. White plastic. Includes seated driver and mechanic figures. "35" decals. Tires read "Oldfield Corp." 1922 winner. Best's original issue of this kit is black plastic with an "8" race number decal. Box art by Rudisill although he signed it "Monaghain."

523 Miller Special (Be): 1959-63, 1/30; $40-50
Ex-Best. White plastic. Includes seated driver and mechanic figures. "23 Bowes Seal Fast Special" decals. Tires read "Firestone." 1931 winner. Box art by Rudisill.

524 Gilmore Special (Be): 1959-63, 1/30; $40-50
Ex-Best. White plastic. Includes seated driver and mechanic figures. "5 Gilmore Speedway" decals. Tires read "Firestone." 1935 winner. Box art by Rudisill.

525 Maserati Special (Be): 1959-63, 1/30; $40-50
Ex-Best. Metallic burgundy plastic. Nicely done two-part spoked wheels. Tires read "Firestone." Includes driver figure. "1 Boyle Spl." decals. 1940 winner. Box art by Rudisill.

526 Fuel Injection Special (Be): 1959-63, 1/30; $40-50
Ex-Best. Light gray plastic. Includes a driver figure not in the original Best issue. "14 Fuel Injection Special" decals. Tires read "Firestone." The 1954 winner. Best's original issue of this car is dark yellow plastic, representing the color of the car when it won in 1953. Box art by Rudisill.

Ex-Advance Cars
All have vinyl tires imprinted with the "Champion" name, but do not have clear or chrome parts.

511 MG (Ad): 1959-69, 1/32; **$30-35**
Ex-Advance. Red plastic. Vinyl tires, no clear or chrome parts. A 1951 MG-TD. Also issued as **590** and **541** Custom. Box art by Cox.

512 Jaguar XK-120 (Ad): 1959-71, 1/32; **$30-35**
Ex-Advance. Ivory plastic. Vinyl tires, no clear or chrome parts. The only ex-Advance model with a one-piece body. The box photo is of an XK-140. Also issued as **580** and **542** Custom.

513 Ferrari America (Ad): 1959-67, 1/32; **$30-35**
Ex-Advance. Dark red plastic. Vinyl tires, no clear or chrome parts. Coupe version of **514**. In first issue box art Jim Cox painted a road sign pointing to Fairton and Cedarville, towns down the road from his home in Bridgeton, symbolized by a covered bridge.

514 Ferrari Sportster (Ad): 1959-67, 1/32; **$30-35**
Ex-Advance. Green and tan plastic. Vinyl tires, no clear or chrome parts. Convertible version of **513**. Also issued as **543** Custom. Box art by Cox.

515 Cunningham (Ad): 1959-67, 1/32; **$30-35**
Ex-Advance. Gray plastic. Vinyl tires, no clear or chrome parts. Also issued as **544** Custom. Box art by Cox.

580 Jaguar (Ad): pre-59, 1/32; **$60-70**
Ex-Advance model in an Advance box with Aurora trademark. Light yellow plastic. "Famous Sports Cars of All Nations." Also issued as **512**. Only the Jaguar and MG are known to exist in this packaging, but the other ex-Advance cars may also have been issued in this series.

590 MG TD (Ad): pre-59, 1/32; **$60-70**
Ex-Advance model in an Advance box with Aurora trademark. Blue plastic. "Famous Sports Cars of All Nations." Also issued as **511** and **541**.

Customized Cars

Four ex-Advance models and six Aurora original sports cars were given added parts, including chrome parts, and a decal sheet with various detailing touches to "customize" his model.

541 Customized MG: 1961-64, 1/32; **$30-35**
Dark red plastic. Black plastic hardtop, chrome parts added. Bullet decal. Also issued as **511, 590**.

542 Customized Jaguar: 1961-64, 1/32; **$30-35**
Ivory plastic. Black hardtop, chrome parts added. Leaping jaguar decal. Also issued as **512, 580**.

543 Customized Ferrari: 1961-64, 1/32; **$30-35**.
Green, beige, chrome plastic. Tiger Shark's mouth decal. Also issued as **514**.

544 Customized Cunningham: 1961-64, 1/32; **$30-35**.
Gray, chrome plastic. Decal of old radio series hero *The Phantom*. Also issued as **515**.

545 Customized Mercedes: 1962-64, 1/32; **$30-35**
Silver, chrome plastic. Lightning decal. Also issued as **517**.

546 Customized Triumph: 1962-64, 1/32; **$30-35**
White plastic. Black hardtop, chrome parts added. Wildcat decal. Also issued as **518**.

547 Customized Corvette: 1962-64, 1/32; **$30-35**
Red, clear plastic. Black, chrome custom parts. "Astrovette" decal. Also issued as **519**.

548 Customized T-Bird: 1962-64, 1/32; **$30-35**
White, clear plastic. Added black headrests, bubble skirts, chrome parts. Decal for "Thunderhawk." Also issued as **520**.

549 Customized Alfa Romeo: 1964, 1/32; **$35-40**
Beige, clear plastic. "Ramjet" decal. Also issued as **510**.

550 Customized Austin-Healey: 1964, 1/32; **$35-40**
Red plastic. Black, chrome custom parts. "Sonic I" decal. Also issued as **516**. Box art by Künstler.

1/32 Sports Cars

Aurora issued 28 original sports car models in this scale. The earliest kits, **516, 517,** and **518,** were made like the Advance cars, without clear or chrome parts but with black vinyl tires. Later issues added clear parts and sometimes chrome parts. Cars issued in 1965 and thereafter have two-part plastic tires in the same plastic as the car.

506 Chevy "Monza" GT: 1965-71, 1/32; **$50-55**
Cream, clear plastic. Last 1/32 car issued with black tires, but they are plastic, not vinyl. Racing decals. 1964 Corvair coupe.

510 Alfa Romeo: 1962-69, 1/32; **$30-35**
Beige, clear plastic. Vinyl tires. Also issued as Customized **549**. Box art by Leynnwood.

516 Austin Healey: 1960-69, 1/32; **$30-35**
Red plastic. Vinyl tires, no clear or chrome parts. 1958 convertible. An unusual kit in this series because it includes a driver figure—a lady driver, at that! Also issued as Customized **550**.

517 Mercedes Benz 300SL: 1960-71, 1/32; **$30-35**
Silver plastic. Vinyl tires, no clear or chrome parts. 1956 Mercedes convertible. An unusual kit in this series because it includes a driver figure. Also issued as Customized **545**.

518 Triumph TR-3: 1960-69, 1/32; **$30-35**
White plastic. Vinyl tires, no clear or chrome parts. Also issued as Customized **546**. The man in the photo on the box is George Burt.

519 Corvette: 1962-71, 1/32; **$30-35**
Red, clear plastic. Vinyl tires. 1961 Corvette convertible. Also issued as Customized **547**. Box art by Leynnwood.

520 Thunderbird: 1962-71, 1/32; **$30-35**
White, clear plastic. Vinyl tires. 1960 Thunderbird. Can be built as hardtop or convertible. Includes a folded-down convertible roof part—but the actual car stored the folded roof in the reverse-opening trunk. Also issued as Customized **548**. Box art by Leynnwood.

533 XKE Gran Turismo: 1963-71, 1/32; **$35-40**
Red, clear plastic. Vinyl tires. 1961 coupe.

538 Triumph Spitfire: 1964-69, 1/32; **$35-40**
Yellow, clear plastic. Vinyl tires. 1963 Triumph.

539 Porsche Carrera: 1964-70, 1/32; **$40-45**.
Green, clear plastic. Vinyl tires. 1958 Porsche coupe.

540 Mustang: 1965-71, 1/32; **$50-55**
Dark red, clear plastic. Black plastic tires. Racing stripe decal. 1964 Ford convertible.

559 Ford GT: 1965-70, 1/32; **$50-55**
White, clear plastic. Racing decals. GT40 Mark I. Body shell used in Aurora and K & B slot cars.

601 Sunbeam Tiger: 1965-69, 1/32; **$35-40**
Beige, clear plastic. Racing decals. 1964 British Sunbeam Alpine modified by Carroll Shelby to take a Ford V-8 engine.

602 Fiat Abarth: 1966-69, 1/32; **$60-65**
Beige, clear plastic. Racing decals. Listed in 1966 catalog and on instruction sheet as **674**.

664 Pontiac GTO: 1965-71, 1/32; **$40-50**
Red, clear, chrome plastic. Five figures in beige plastic. Racing decals. Body shell used in Aurora and K & B slot cars. 1965 Pontiac. Reissued by Monogram.

665 Ford Mustang Fastback: 1965-71, 1/32; **$30-40**
Cream, clear, chrome plastic. 1965 Mustang. Body shell used in Aurora and K & B slot cars. Box art by Dunham. Reissued by Monogram.

666 Corvair Corsa: 1967-70, 1/32; **$40-50**
Light yellow, clear, chrome plastic. This model reverts to Advance-style wheels attached to studs on chassis. 1965 Corvair sportster. Body shell used in Aurora and K & B slot cars.

667 Plymouth Barracuda: 1967-70, 1/32; **$40-50**
Red, clear, chrome plastic. Racing decals. Later $1 issue has five figures in beige plastic. 1965 Barracuda. Same tooling used in **680 Hemi Under Glass**. Body shell used in Aurora and K & B slot cars. Reissued by Monogram.

668 Chevy-Chaparral: 1966-71, 1/32; **$35-45**
White, clear, chrome plastic. Racing decals. Later $1 kit has five figures in beige plastic. Body shell used in Aurora and K & B slot cars.

669 Cobra GT Coupe: 1966-71, 1/32; **$40-50**
Blue, clear, chrome plastic. Racing decals. 1964 Cobra Daytona. Body shell used in Aurora and K & B slot cars.

670 Comet Exterminator: 1967-70, 1/32; **$40-50**
Red, clear, chrome plastic. Hurst racing decals. 1964 Mercury funny car matched with **680**. Has small drag front wheels, injector scoop on hood, and dragster interior with just driver's bucket seat and roll bar. This is a model of Sachs & Sons A/FX dragster (*Hot Rod*, September 1964). Body shell used in Aurora and K & B slot cars. Box art by Dunham.

671 Lola T-70: 1966-71, 1/32; **$40-50**.
Blue, clear, chrome plastic. Racing decals. Later $1 issue has five beige figures. Decals. The Lola/Ford Zerex Special. Body shell used in Aurora and K & B slot cars.

672 Rover BRM: 1968-71, 1/32; **$50-60**
Metallic green, clear, chrome plastic. Racing decals. Five beige figures. Gas turbine Le Mans GT racer. Body shell used in Aurora and K & B slot cars.

673 Demolition Demon: 1966-70, 1/32; **$80-90**
Medium blue, clear plastic. Derby decals. A badly dented demolition derby 1957 Ford! A difficult pattern for HMS to make, but part fit was good. Box art by Dunham.

675 Chevy Monza SS: 1965-70, 1/32; **$50-60**
Red, clear, chrome plastic. 1964 Corvair convertible. Listed in 1965, 1966 catalogs as **602**.

677 Mako Shark: 1968-71, 1/32; **$50-60**
Metallic blue, clear, chrome plastic. Five beige figures. L88 Corvette. Reissued by Monogram.

680 Hemi Under Glass: 1967-71, 1/32; **$70-80**
Black, clear, chrome plastic. "Hurst" racing decals. George Hurst funny car. Matched with **670**. Features chrome Chrysler "hemi" engine under fastback rear window. Same body mold as **667**.

681 Hurst Baja Boot: 1970-71, 1/32; **$50-60**
Bronze plastic, clear, chrome. Racing decals. Five beige plastic figures. George Hurst off road racer that won the 1969 Baja 500. Reissued as **620** Dune Buggy in 1976 hot rod series.

Original Issue 1/32 Hot Rods

Aurora made 22 hot rod models in this scale. None of the hot rods has black vinyl tires or chrome parts, but all have clear parts and a decal sheet with nicknames and body trim. All but **603** use the same Chevy 409 engine; however, carburetors, manifolds, and exhaust pipes are individualized.

507 1927 Model T Ford Sad Sack: 1962-71, 1/32; **$45-55**
Yellow, clear plastic. Includes parts for standard or oversize racing slick rear tires; optional "moon disc" or "embossed" hubcaps that snap in and out of the wheel rims. This model closely resembles the car of New Mexico's Keith Heiskell (*Hot Rod Yearbook*, No. 1, 1961).

508 1929 Ford A Pick Up Wolf Wagon: 1962-70, 1/32; **$50-60**
Red, clear plastic. Pick-up truck. Standard or racing slick rear tires; optional "moon disc" or "embossed" hubcaps. This car seems to be modeled after Ken Kay's show rod (*Hot Rod*, July 1961).

509 '32 Ford Ram Rod: 1962-71, 1/32; **$45-55**
Blue, clear plastic. Three-window coupe. Standard or racing slick rear tires; optional "moon disc" or "embossed" hubcaps. This is a model of Clarence Catallo's "Silver Sapphire" that appears of the cover of the Beach Boys' 1963 *Little Deuce Coupe* record album (*Hot Rod*, July 1961).

527 '21 Ford T for Two: 1963-71, 1/32; **$45-55**
Black, clear plastic. Coupe. Standard or racing slick rear tires; optional "moon disc" or spoked wheels. This is a model of Joe Cruces' show rod (*Rod and Custom*, October 1962). Box art by Künstler.

528 '27 Ford T Shiftin Drifter: 1963-71, 1/32; **$45-55**
Cream, clear plastic. Model T pickup with removable "canvas" bed cover. Standard or racing slick rear tires; optional "moon disc" or spoked wheels. Reissued as **625**.

529 '29 Ford Pick Up Beatnik Box: 1963-71, 1/32; **$45-55**
Yellow, clear plastic. Pick-up truck. Standard or racing slick rear tires.

530 Show Trailer: 1963-66, 1/32; **$20-25**
Light blue plastic. Part of a set with **555 Chevy Pick-Up** and **554 32 Skid Doo**.

534 1934 Ford Rolls Scat Cat: 1964-71, 1/32; **$45-55**
Red, clear plastic. Has a Rolls-Royce radiator. Racing slicks in rear. Box art by Künstler.

535 Ford T Dragster Spyder: 1964-71, 1/32; **$45-55**
Yellow, clear plastic. Optional embossed or spoked wheel covers. Model of the car built by Atlanta Speedshop-Dragmaster and later modified by Higley & Hubbard of California (*Hot Rod*, September 1964). Reissued as **623**. Box art by Künstler.

536 1923 Ford T Rod The Charger: 1964-69, 1/32; **$45-55**
Bronze, clear plastic. Reissued as **597, 629**. Racing slicks in rear. Box art by Künstler.

537 '28 Chevy Moody Monster: 1964-70, 1/32; **$60-70**
Red, clear plastic. Roadster. Reissued as **627**. Racing slicks in rear. This is a model of Hugh Tucker's car (*Hot Rod*, August 1963). Box art by Künstler.

551 '29 Ford "Touring" Road Raider: 1965-68, 1/32; **$50-60**
Red, clear plastic. Reissued as **594**. Box art by Künstler.

552 '30 Ford Fire Engine Hot Surfer: 1965-68, 1/32; **$70-75**
Red, clear plastic. Model A fire truck with surf boards in back. Reissued as **596**. Box art by Künstler.

553 '39 La Salle Hearse with a Curse: 1965-67, 1/32; **$75-80**
Black, clear plastic with engraved curtains on side windows. Surf wagon. Unusual subject and short production life make this a high priced kit for collectors. Box art by Künstler.

554 '32 Ford 32 Skid Doo: 1963-67, 1/32; **$45-55**
Red, clear plastic. Optional "moon disc" or "embossed" hub caps. Part of a set with **530, 555**. Reissued as **592, 621**. This car was inspired by the 1960 version of Dave Stuckey's "Lil' Coffin" (*Car Craft*, November 1960). After Stuckey further customized the car, Monogram turned it into its "Lil' Coffin" model.

555 Chevy Custom Pick-Up Draggin Wagon: 1963-69, 1/32; $45-55
Red, clear plastic. A 1962 Chevrolet. Part of a set with **530, 554**.

556 '24 Get Away Buick: 1964-68, 1/32; **$45-55**
Black, clear plastic. Soft-top sedan with roll bar. Spoked wheels. Hood lifts off to show motor. Reissued as **624**. This is a model of Jim Morris' car (*Hot Rod*, August 1963). Box art by Künstler.

557 '22 Ford Hi Stepper: 1965-68, 1/32; **$50-60**
Black, clear plastic. Racing slicks in rear. Reissued as **591, 622**. Inspired by Southern Californian Jim Spurbeck's T highboy (*Hot Rod*, August 1963). Box art by Künstler.

558 '30 Ford Woodin Wagon: 1965-68, 1/32; **$50-60**
Brown, clear plastic. Like Jan and Dean's "Surf City" station wagon, this "woodie" "doesn't have a back seat or a rear window." Does have a sun roof. Reissued as **593, 628**. Box art by Künstler.

603 '27 Ford T Coupe Snap Dragin: 1966-69, 1/32; **$50-60**
Black, clear plastic. Quad over-under headlights. Small spoked bicycle wheels in front, racing slicks in rear. Bucket racing seats. This was the last kit to be released in the original hot rod line and the only one that did not use Aurora's Chevy engine. It has a Ford 390 V-8.

662 '37 Packard Ambulance Meat Wagon 1965-68, 1/32; **$85-90**
White, clear plastic. Collectors seek this because of unusual subject and short production life. Box art by Künstler.

663 Old Ironsides: 1965-67, 1/32; **$65-75**.
Metallic green, clear plastic. "The Tank" decal. This kit is identified as "The Armored Car" in the Aurora catalog. Reissued as **595**. Box art by Künstler.

Scene Machine 1/32 Hot Rods

In 1970 Aurora modified seven of its hot rod models into "hippy" vehicles. All have clear and chrome parts. Each has a decal sheet with two decorative versions: "Way Out" or "Real Cool." A standing mod couple came with each kit.

591 Drop Out Bus: 1970-72, 1/32; **$35-45**
Yellow, clear, chrome plastic. "Real Cool" version is a school bus motif. Also issued as **557, 622**.

592 Peppermint Fuzz: 1970-71, 1/32; **$40-50**
Pink, clear, chrome plastic, metal chain. Added: machine gun on roof, star on grille, "Hog Chopper" motorcycle on running board. Also issued as **554, 621**.

593 Boob Tube: 1970-72, 1/32; **$35-45**
Also issued as **558, 628**.

594 Black Beard's Tub: 1970-72, 1/32; **$35-45**
Bronze, clear, chrome plastic, metal chain. Added: surf board, anchor, ice chest, sand dune, and palm tree. Reissue of **551**.

595 Butterfly Catcher: 1971-72, 1/32; **$40-50**
Lime green, transparent dark green, chrome plastic. Reissue of **663**.

596 The Wurst: 1970-72, 1/32; **$35-45**
Red, clear, gold chrome plastic. Hot dog stand on old fire truck. White or yellow plastic umbrella and kitchen equipment added. Reissue of **552**.

597 Teepee T 1970-72, 1/32; **$35-45**
Blue, tan, clear, chrome plastic. "Way Out" flower power decals or "Real Cool" Native American art. Also issued as **536, 629**.

Final Issue Hot Rods

After being out of the catalog for three years, ten hot rod models were revised for a final issue. All have clear and chrome parts, fat "Indy Profile" tires, and new decals.

620 Dune Buggy: 1976-77, 1/32; **$25-30**
Yellow, clear, chrome plastic. Roof and spare tires deleted from earlier Baja Boot version. Reissue of **681**.

621 32 Ford Sedan 1976-77, 1/32; **$25-30**
Red, clear, chrome plastic. Because the 32 Ford is *the* classic hot rod, this kit is the hardest in this series to find today. Reissue of **554, 592**.

622 22 T Sedan 1976-77, 1/32; **$20-25**
Black, clear, chrome plastic. Reissue of **557, 591**.

623 T Dragster: 1976-77, 1/32; **$20-25**
Red, clear, chrome plastic. Frame lengthened to extend front wheels in dragster configuration. Red, white, and blue color scheme for the American Revolution Bicentennial. Reissue of **535**.

624 24 Buick Touring: 1976-77, 1/32; **$20-25**
White, clear, chrome plastic. Reissue of **556**.

625 32 Ford Pickup: 1976-77, 1/32; **$20-25**
Blue, clear, chrome plastic. Reissue of **528**.

626 21 T Coupe: 1977, 1/32; **$20-25**
Dark red, clear, chrome plastic. Reissue of **527**.

627 28 Chevy Roadster: 1977, 1/32; **$20-25**
Black, clear, chrome plastic. Reissue of **537**.

628 29 Ford Woodie: 1977, 1/32; **$20-25**
Light blue, clear, chrome plastic. Reissue of **558, 593**.

629 1912 T Truck: 1977, 1/32; **$20-25**
Green, clear, chrome plastic. Reissue of **536, 597**.

1/32 Fire Truck

599 American La France 900 Pumper: 1964-68, 1971-72, 1/32; **$50-70**
Red, black, silver, cream, clear, and chrome plastic. Box art by Steel. Reissued in 1971-72 as **California Fire Truck** in white plastic with chrome wheels. California issue comes in a square, white box with art by John Amendola and sells in the **$70-90** range.

1/25 Scale Cars

The 1/25 cars have clear and chrome parts and vinyl-rubber tires. All have hoods, trunks, and doors that open. Five of the sports cars were reissued in 1971-72 as the "Battle Aces of the Road" (**577-581**) with new numbers and special decals.

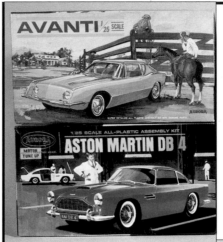

560 Studebaker Avanti: 1964-67, 1/25; **$30-40**
Beige, black, clear, chrome plastic. 1963 Avanti coupe. Studebaker had stopped production of the Avanti before Aurora's model reached the hobby shops! Hood, doors, and trunk opens to show luggage inside.

561 Porsche 904 GT: 1965-70, 1/25; **$20-30**
Gray, black, clear, chrome plastic. Porsche supplied Aurora with blueprints and photos for this model. *Scale Auto* (April 1992) said: "The Aurora model was a very complex kit which featured working steering, opening engine cover and opening front compartment. The one major drawback of this model was that although it had lots of separate parts, none of them were engraved to any particularly great level of detail." Reissued as **578**. Reissued by Monogram.

562 Aston-Martin DB-4: 1965-68, 1/25; **$20-30**
White, red, black, clear, chrome plastic. 1964 coupe. Doors, hood, trunk open. Pattern reworked and issued as **585**. Reissued by Monogram.

563 Ferrari Berlinetta: 1964-70, 1/25; **$20-30**
Red, black, clear, chrome plastic. Doors, hood open; side windows slide open. 1962 Ferrari 250-GTO. Aurora began working on the pattern for a 1961 GT, but when Ferrari brought out the 250-GTO, Aurora changed to the GTO body but kept the GT chassis, making the

body a little "stubby" to fit on the shorter GT chassis. Also, because this model used the same wheels as the Jaguar XKE, they are slightly off scale. Reissued as **579** and by Monogram. Box art by Leynnwood.

564 Maserati 3500: GT 1964-67, 1/25; **$20-30**
Gray, red, clear, chrome plastic. 1963 coupe. Doors, hood, trunk open. Reissued by Monogram. Box art by Leynnwood.

565 Ford GT: 1965-70, 1/25; **$20-30**
White, black, clear, and chrome plastic. Racing decals. Rear engine hood opens, front spare tire compartment opens. GT40 Mark I. Body shell used in K & B slot car. *Model Car and Track* (April 1965) said: "The body of the Ford GT was apparently modeled from early releases before the full-sized design was finalized, with the result that the tail and the underside of the nose do not conform to any car actually built." Reissued as **580**. Box art by Dunham.

566 XKE Jag: 1962-70, 1/25; **$20-30**
Red, clear, chrome plastic. Coupe. Hood opens. Reissued as **577** and by Monogram. Box art by Leynnwood.

567 Jaguar XKE Convertible: 1963-70, 1/25; **$25-35**
White, black, clear, chrome plastic. Hood opens. Contains extra parts for hard top version. Box art by Leynnwood.

568 Double Deuce '22 Ford: 1963-67, 1/25; **$60-80**
Kit contains Roadster and Dragster 1922 Model T's produced from two separate molds. Roadster in black, clear, and brass plated plastic.

Hood lifts off to show engine. Dragster in metallic red, clear, chrome plated plastic. Box art by R. Schulz.

569 '34 Ford Stock and Street Rod: 1963-67, 1/25; **$130-150**
Kit contains models of two 34 Fords, produced from separate molds. Stock coupe in black, clear, chrome plastic. Street Rod in metallic green, clear, chrome plastic. "Scream Puff" decals. Monogram reportedly still has the molds. Box art by R. Schulz.

570 Undertaker Dragster: 1964-67, 1971-72, 1/25; **$70-90**
Black, white, clear and chrome plastic. Carl Casper's dragster, winner of best show car at 1963 National Hot Rod Association's Championship. Includes base, tombstone, figures of ghost and undertaker. 1971 reissue comes in square, white box and sells in the **150-160** range. Black and lime green plastic.

577 XKE Jag Spitfire: 1971-72, 1/25; **$40-50**
Battle Aces reissue of **566**. Orange, clear, chrome plastic. Front hood opens. British World War II RAF aircraft inspired decals.

578 Porsche Messerschmitt: 1971-72, 1/25; **$60-70**
Battle Aces reissue of **561**. Black, clear, chrome plastic. German World War II Luftwaffe aircraft inspired decals.

579 Ferrari Scorpion: 1971-72, 1/25; **$60-70**
Battle Aces reissue of **563**. Red, black, clear, chrome plastic. Italian World War I air corps inspired decals.

580 Ford GT Flying Tiger: 1971-72, 1/25; **$60-70**
Battle Aces reissue of **565**. Green, black, clear, chrome plastic. United States World War II Flying Tiger aircraft inspired decals.

581 Chaparral Lightning: 1971-72, 1/25; **$70-80**
Battle Aces reissue of **584**. Silver, black, clear, chrome plastic. United States World War II Air Force inspired decals.

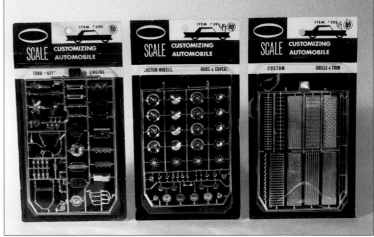

582 Archie's Car: 1969-71, 1/25; **$80-100**
Yellow and clear plastic. Hood lifts off to show engine. Lumpy "wacky action" plastic wheels. Based on the Archie comic book and TV cartoon show characters. Figures for Archie, Veronica, Hot Dog.

590 Ford 427 Engine: 1964, 1/25; **$15-20**
Chrome. 46 parts blister packed on a card. In the mid 1960s Revell issued Custom Car Parts kits to help modelers in the Revell national car customizing contests—so Aurora did too.

591 Customizing Automobile Parts: 1964, 1/25; **$10-15**
Chrome wheels, Hub caps. Blister packed on a card.

592 Customizing Automobile Parts 64, 1/25; **$10-12**
Chrome grills and trim. Blister packed on a card. The grills are molded in rectangular strips so that modelers can trim them to fit any grill shape.

583 Mod Squad Station Wagon: 1970-71, 1/25; **$100-120**
Black, clear, and chrome plastic. Hood and doors open. Stripe decals for surfboards. Figures and surfboards beige. Box says it is a 1951 Mercury and the instruction sheet says 1949. Based on TV show. Contains figures for Link, Pete, and Julie. Last new TV character kit produced by Aurora. Production delayed until second season of show; then this woodie wagon was driven off a cliff in second episode! Although mold was beryllium, it was an expensive kit to produce.

584 Chevy-Chaparral: 1966-70, 1/25; **$80-90**
White, dark brown, clear, chrome plastic. #3/Firestone racing decals. Jim Hall Chevy-powered Chaparral II GT. Said model developer Ray Haines: "We had a hell of a time getting information from Hall on it because everything was under wraps." Last of the original 1/25 sports cars series—and the most rare kit of the series today. Monogram reportedly has the mold, minus the wheels tooling.

585 Aston Martin Super Spy Car: 1967-68, 1/25; **$120-140**
Gray, clear, chrome plastic. Two seated figures. Based on **562** pattern. An Aston Martin DB5. Although James Bond's name is not on this kit, it is shown in the catalog across from the Bond figure kit. (Airfix held the license to the official Bond car kit.) Battering ram bumpers, front turn signal machine guns, revolving license plates, passenger ejection seat, hubcap tire cutters, rear movable bullet shield, and tail light oil slick and smokescreen spouts.

832 Banana Buggy: 1969-71, 1/25; **$600-700**
Yellow plastic. From the Hanna-Barbera TV show *Banana Splits*. Buggy has optional animal tail parts to snap on to back and decal selections to customize buggy for each of the four vehicles driven by the TV characters: a lion, dog, elephant, and ape. Includes four Banana Splits figures which can be placed in driver's seat or on a park bench. This model was created for the under-10 market at a time when Aurora was moving into the toy field.

1/16 Scale Old Timers and American Classics
The first three Old Timers came out in 1961, followed by another three in 1963. The kits were Aurora's most detailed auto models at the time. In 1975 they were revised and reissued as American Classics.

571 Stutz Bearcat: 1961-67, 1/16; **$20-30**
Yellow, brown, clear, chrome plastic. Rubber tires. 1914 two-seat roadster. No engine parts under the hood. A model of Hudson Miniatures president Anthony Koveleski's car. Reissued as **156**.

572 Mercer Raceabout: 1961-67, 1/16; **$20-30**
Black, white, clear, chrome plastic. Rubber tires. 1911. No engine parts under the hood. Reissued as **155**.

573 Stanley Steamer: 1961-67, 1/16; **$20-30**
Red, black, clear, chrome plastic. Rubber tires. Bent metal hangers to attach fenders. 1909. Reissued as **154**.

574 Buick Bug: 1963-67, 1/16; **$20-30**
Green, ivory, clear, chrome plastic. Rubber tires. 1911. Hood lifts off to show 14 horsepower engine. Reissued as **153**.

575 Rambler: 1963-67, 1/16; **$20-30**
Yellow, black, clear, chrome plastic. Rubber tires. 1903. Driver's seat lifts up to show one-cylinder engine. Reissued as **152**.

576 Curved Dash Olds: 1963-67, 1/16; **$20-30**
Red, black, clear, chrome plastic. Rubber tires. 1904. Single-cylinder engine under seat; chain drive. Reissued as **151**.

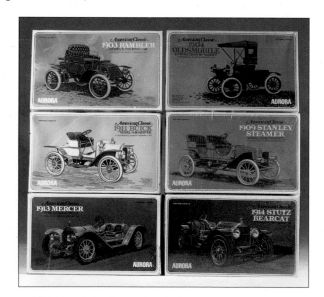

151: 1903 Rambler Model E: 1975-77, 1/16; **$20-30**
Red, black, clear chrome plastic. White rubber tires. Reissue of **575**. Box art by Amendola.

152 1904 Oldsmobile Curved Dash Runabout: 1975-77, 1/16; **$20-30**
Black, metallic gray, clear, chrome plastic. White rubber tires. Reissue of **576**. Box art by Amendola.

153 1911 Buick Model 13 Roadster: 1975-77, 1/16; **$20-30**
Green, white, clear, chrome plastic. White rubber tires. Reissue of **574**. Box art by Amendola.

154 1909 Stanley Steamer Model E2 Runabout: 1975-77, 1/16; **$20-30**
Yellow, green, clear, chrome plastic. White rubber tires. Floorboard parts with slots for foot pedals deleted. Replaced by textured areas on chassis part and holes for pedals. Metal parts in original issue replaced with plastic. Reissue of **573**. Box art by Amendola.

155 1913 Mercer Type 35J Raceabout: 1975-77, 1/16; **$20-30**
Yellow, clear, chrome plastic. Rubber tires. Reissue of **572**. Box art by Amendola.

156 Stutz Bearcat: 1975-77, 1/16; **$25-35**
Red, black, clear, chrome plastic. Rubber tires. Reissue of **571**. Box art by Amendola.

1/16 Scale Racing Scenes

In 1974 Aurora issued this set of eight very detailed and expensive funny car component kits. All of the car parts and figures could be integrated into "Hank's Speed Shop" garage diorama. In 1975 the engine, body, chassis and driver, kits were combined to create two "Super Scale" car kits: **851, 852**.

841 Four Funny Car Drivers 74, 1/16; **$20-25**
Contains the same two driver figures, molded once in silver and again in orange plastic—the colors of driver firesuits. Heads and arms are movable and interchangeable. Includes six racing helmets, two trophies, and decals for uniforms.

842 Racing Mechanic With Tools 74, 1/16; **$20-25**
Bright orange plastic. Head and arms interchangeable with other figures in set. Includes large tool chest, small tool box, tools, under-car creeper, welding tanks, and torch.

843 Donovan 417 Blown Engine 74, 1/16; **$50-55**
Metallic gray and chrome plastic. Pistons and crankshaft move. Wires for electrical system, plastic tube for cooling system. Decal sheet.

844 Chrysler 392 Fuel Injected Engine 74, 1/16; **$45-50**
Metallic gray and chrome plastic. Pistons and crankshaft move. Wires for electrical system. Decal sheet.

845 Vega Funny Car Body and Custom Painters 74, 1/16; **$45-50**
Yellow body, 2 blue plastic figures with movable parts that interchange with other figures in the series. Clear and chrome parts. One piece body snaps onto pivots at rear of chassis kit **847**. "Viking" decal painting done by Nat Quick, painter of real dragsters. It is noteworthy that one of the painters is black, a rarity in the nearly all-white world of plastic models. Of this, Jim Keeler said, "Nobody else had done anything like that, and I thought it was high time."

846 Pinto Funny Car Body and Custom Painters 74, 1/16; **$45-50**
Blue plastic one piece body, clear, chrome, 2 beige figures. "Mustang" decal.

847 Funny Car Chassis 74, 1/16; **$90-100**
Black, chrome plastic. Includes two optional sets of wheel rims and vinyl tires. Front wheels steer. Any combination of engine and body snaps on to the chassis.

851 Super Scale Vega Funny Car 75, 1/16; **$210-230**
Body molded in either yellow or black. Decals for *Intruder* or *Voodoo*, Aurora kit director Tom West's real car. Includes the same features as **852**.

852 Super Scale Pinto Funny Car 75, 1/16; **$210-230**
Body molded in Blue, yellow, or orange. Clear, chrome, black, metallic gray, silver plastic. Vinyl tires. Wire and rubber tube for wiring and cooling systems. Two driver figures. Donovan engine. 260 parts. Decals for *Phantom* or *Magician*.

Motorized Cars
In 1961 Aurora joined a fad popular among model companies at the time by issuing two battery powered car kits. Power came from two D batteries. Although large scale, the kits were poor in accuracy and detail.

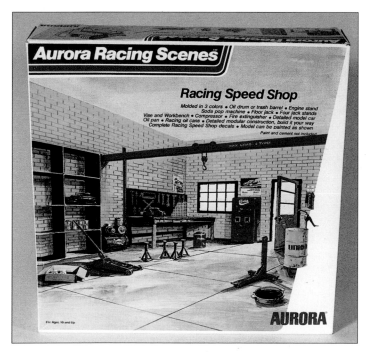

848 Racing Speed Shop 74, 1/16; **$130-150**
Red, yellow, cream plastic. Floor, two walls, and garage equipment. Includes small Model A car model on shop shelf. Named for *Car Model* magazine writer Hank Borger, recently badly burned in a garage fire. Two wall calendars feature names of kit's creators—"T. West Racing Graphics" and "Ideas Unlimited: Keeler and Yanchus."

531 Customized Motorized Pontiac: 1961, 1/11; **$70-80**.
White, clear, transparent red, chrome plastic. Black plastic tires. Decals. 1959 Pontiac Bonneville. Motorized version of **565**. Includes two working light bulbs for headlights, motor, rubber drive belt to turn rear wheels.

137

532 1932 Ford Hotrod: 1961-66, 1/11; **$60-70**
Red and clear plastic. Black plastic tires. "Fire Ball" decals. Includes two working light bulbs for headlights, motor, rubber drive belt to turn rear wheels.

565 Pontiac Fireball: 1961-63, 1/11; **$60-70**
White, black, clear, transparent red, chrome plastic. Decals. Same kit as **531** without motor.

Leapin Lena 61, 1/32; **$100-110**
Yellow plastic with black plastic wheels. A 1930s hot rod. Includes an electric motor that is mounted as a sidewinder over the rear axle. A friction drive gear turns the right rear wheel. Powered by a size C battery. This "Go Toys" item is a companion to the motorized car kits (**531, 532**).

Science Fiction

129 Nuclear Airliner: 1960-64, 1/200; **$90-100**
White, clear plastic. "Impetus." Based on speculative articles in popular science magazines. Rear of fuselage opens to show nuclear engine. Rocket bay swings open. Includes small booster aircraft. Reissued as **251**.

148 Pan Am Space Clipper: 2001: 1969-70, 1/144; **$100-120**
White, clear plastic. On standard clear aircraft stand. From the movie *2001 A Space Odyssey*. Aurora designed model from plans supplied by MGM, but special effects model used in movie differs in details because it departed from the same plans. Rear of fuselage removes to show nuclear engine. Reissued as **252**.

251 Ragnarok Orbital Interceptor: 1976-77, 1/200; **$90-100**
Black, transparent amber plastic. Revised issue of **129**. Booster craft's nose boom removed and hole in rear of Interceptor's fuselage enlarged so booster can park. Decal locater lines removed.

252 Space Shuttle Orion 1976-77, 1/144; **$100-120**
White, clear plastic. On standard aircraft stand molded in black plastic. Revised issue of **148**. Textured panels on original model are made smooth.

253 Nuclear Submarine Seaview: 1976-77, 1/316; **$180-200**
Gray, clear plastic. Radar mast rotates. Revised issue of **707**. Changes for reissue: uses base from **Sealab** model (light green plastic), hull plastic color changed from black to gray, raised panel lines added to hull.

254 Flying Sub 1976-77, 1/60; **$90-100**
Yellow, blue, clear plastic. Revised issue of **817**. Interior plastic color changed from silver to metallic blue. Reissued by Monogram, Tsukuda. Box art by Schaare.

255 Rocket Transport Spindrift: 1976-77, 1/64; **$100-120**
Orange, light green, transparent red, clear plastic. Decal. Regular aircraft stand in black plastic. Reissue of **830** with correct red dome replacing clear dome and raised lines for fuselage insignia replaced by a decal.

256 Flying Saucer: 1976-77, 1/72; **$70-90**
Gray plastic. Reissue of **813**. Changes: base (brown plastic) and three figures (Tracy, Junior, and Moon Maid) from **Space Coupe 819** used, underside globes changed from silver to transparent red, optional clear top eliminated. Reissued by Monogram, Tsukuda.

418 The Robot: 1968-70, 1/11; **$400-500**
Silver, clear plastic. From 1965-68 TV show *Lost in Space*. Electric bolt flashes from right hand.
Box art by Schaare.

419 Lost in Space: 1967-70, 1/32; **$300-350**
Tan plastic. One-eyed cyclops on small cliff threatening five Robinsons. Inscription on base reads, "Lost in Space." Pattern by Meyers. Box art by Künstler.

420 Lost in Space: 1967-70, 1/32; **$400-450**
Tan, gray plastic. Decals for chariot's windows. Same kit parts as **419** with added rock parts to make taller cliff and "chariot" tracked vehicle added. Box art by Künstler.

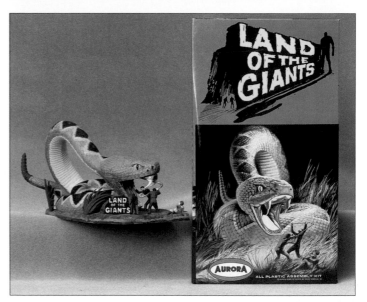

486 Batmobile: 1967-70, 1/32; **$250-300**
Black, chrome, clear plastic. Bat decals. From *Batman* TV show. First issue box is purple, second is light blue.

487 Batplane: 1967-70, 1/60; **$200-250**
Black and clear plastic. Bat decals. Simple model with few parts, minimal detail. Landing gear, but no gear wells. Batman and Robin heads in sparse cockpit. Mounted on regular aircraft stand decorated with bat wing decals. From *Batman* TV show. Box art by Grinnell.

489 Green Hornet Black Beauty: 1967-69, 1/32; **$150-170**
Black, chrome, clear plastic. Green Hornet decal. Based on the car in the short lived TV series. Andy Yanchus, who sculpted Aurora's HO slot car version of the Black Beauty, criticized the HMS-designed model kit: "The front end was too rounded, and detail was generally mushy."

707 Seaview: 1966-73, 1/316; **$150-200**
Black, clear plastic. Radar mast rotates. Based on 1961 movie and 1964-68 TV show *Voyage to the Bottom of the Sea*. Mounted on sea floor base with nameplate reading "Seaview." Reissued as **253**. Mold has been destroyed. Box art by Grinnell.

810 Bat Cycle: 1968-70, 1/19; **$400-450**
Metallic blue, clear plastic. Bat decals. Based on *Batman* TV show. Customized from a Yamaha Catalina 250. Batman figure on cycle, with Robin on go-cart side car that can be detached.

811 Bat Boat: 1968-70, 1/32; **$400-450**
Metallic blue, clear plastic. Bat decals. Rests on oval base with "Bat Boat" nameplate. Jet powered. Based on *Batman* TV show. Hull was based on V-174 sports boat made by Glastron Boats of Austin, Texas.

813 UFO 1968-71, 1/72; **$80-100**
Silver, clear plastic. Optional clear top. From *The Invaders* TV show. Includes 6 crew figures. Landing struts can be assembled in flying or landing positions. Reissued as **256 Flying Saucer.** Reissued by Monogram, Tsukuda. Box art by Schaare.

816 Land of the Giants: 1969-71, 1/48; **$250-300**
Metallic green plastic. Giant rattlesnake threatens three passengers from lost spacecraft. Nameplate reads "Land of the Giants." Based on 1968-70 TV show *Land of the Giants*. Box art by Schaare.

817 Flying Sub: 1968-70, 1/60; **$130-150**
Yellow, silver plastic. From *Voyage to the Bottom of the Sea*. Scout craft from the Seaview. Top lifts off to reveal two crew members. Pickup hook on bottom of sub raises and lowers. Model is on a clear plastic aircraft stand. Reissued as **254**. Reissued by Monogram, Tsukuda.

819 Dick Tracy Space Coupe: 1968-70, 1/72; **$100-120**
Yellow plastic. Magnetic thrusters swivel. Decal for side reads, "Police, Magnetic Space Coupe." Includes lunar landscape base and four figures: Diet Smith, Tracy, Junior, Moon Maid. Box art by Schaare.

828 Chitty Chitty Bang Bang: 1969-70, 1/25; **$180-200**
Brown, orange or yellow, clear, chrome plastic. From the movie of the same name. Helicopter rotors spin.

829 The Moon Bus: 1969-73, 1/55; **$150-200**
Gray, light green, clear plastic. Decals. From *2001 A Space Odyssey*. Roof lifts off to show four crewmen. No base.

830 Land of the Giants Spaceship: 1969-71, 1/64; **$160-200**
Orange, lime green. Incorrect clear plastic dome atop fuselage. Raised lines on fuselage for insignia. On regular clear plastic aircraft stand. Reissued as **255 Spindrift**. Door slides open, roof lifts off to show crew: pilot Steve Burton, co-pilot Dan Erickson, and stewardess Betty Hamilton.

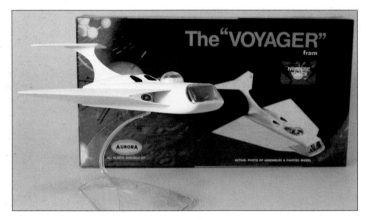

831 The Voyager: 1969-71, 1/96; **$400-450**
White, clear plastic. Very simple kit with no surface detail or decals to enhance its plain surface. Three crew figures. From the movie and TV cartoon show *Fantastic Voyage*. Displayed on an aircraft stand.

921 Enterprise: 1967-73 Canada, 1970-73 England, Holland 1/635; **$250-300**
White or blue plastic; translucent blue and amber plastic. Black plastic stand. Rarest version is the Canada issue with lights in an oval logo box. From the TV show *Star Trek*. This kit is an AMT model, sold under the Aurora trademark in Europe and Canada.

922 Mr. Spock 72 England 1/12; **$200-220**
Black plastic. From *Star Trek*. Spock with three headed snake. AMT issued it in the US in 1973 (white plastic) and in 1979 reissued Spock with a new uniform and no snake. Pattern by Lemon.

923 Klingon Battle Cruiser: 1970-73 England, Holland 1/635; **$200-220**
Gray or blue and transparent green plastic. Comes with black plastic stand. From *Star Trek*. An AMT model packaged and sold by Aurora in Canada and Europe.

Figure Kits

Note on Sculptors: Both Bill Lemon and Ray Meyers were interviewed for this book. Both remembered well most of the model patterns they created for Aurora, and both admitted that they had forgotten all that they did for Aurora. Only those models definitely claimed by Lemon and Meyers are credited to them. The widow of Larry Ehling tried to identify models sculpted by her late husband, but she admitted that in some cases she could not be sure. Thus attributions of models to Ehling are less firm than those for Lemon and Meyers.

The Knights
The Knights were the first figure kits issued by Aurora. The Knights were issued in three sets: the first set was in the catalog continuously from 1956-71 (although catalog numbers changed in 1964), the Camelot Knights came out in 1968, and the metal plated Knights in Shining Armour were made in England in 1973.

K-1 Silver Knight of Augsburg: 1956-63, 1/8; **$40-50**
Silver plastic. Real feather for helmet. Holds sword, helmet visor lifts. This is the first figure kit by any company. Pattern by Lemon. Reissued as **471, 825, 881**.

K-2 Blue Knight of Milan 1957-63, 1/8; **$40-50**
Blue plastic. Real feather for helmet. Holds halberd. Issued with two box art variations: the first has knight standing on a road, the second standing on a drawbridge. Pattern by Lemon. Reissued as **472, 826, 882**.

K-5 Gold Knight of Nice: 1959-63, 1/8; **$250-300**
Gold and black plastic. Based on a 15th century German knight in the Wallace Collection museum, London. Cover story in *Scale Modeler* (July 1966). Pattern by Lemon. Reissued as **475, 885**. Horse also used in Black Fury **400**, Apache Warrior **401**, Confederate Raider **402**.

K-3 Black Knight of Nurnberg: 1957-63, 1/8; **$40-50**.
Black plastic. Real feather for helmet. Holds lance and mace. Helmet face guard swings open. Issued with two box art variations: in the first he stands before a tent with a mace in his right hand, in the second he holds his lance in his right hand and mace in his left. Pattern by Lemon. Reissued as **473, 827, 883**.

207 3 Knights: 1958-59, 1/8; **$400-450**
Includes **K-1, K-2, K-3**. Cement, brush, and paint included.

The 471-475 kits were not modified, only the catalog numbers and box art changed. West Hempstead issued kits in heavy cardboard boxes are more valuable than England issue kits in thin cardboard.

471 Silver Knight: 1964-71, 1/8; **$15-25**
Also issued as **K-1, 825, 881**. Box art by Künstler. The model for the maiden in the cone hat was Künstler's wife.

472 Blue Knight: 1964-71, 1/8; **$15-25**
Also issued as **K-2, 826, 882**. Box art by Künstler.

473 Black Knight: 1964-71, 1/8; **$15-25**
Also issued as **K-3, 827, 883**. Box art by Künstler.

K-4 Red Knight of Vienna 1957-63, 1/8; **$70-80**
Red plastic. Holds lance, helmet lifts off. This, and the Green Knight pattern, were the only knight figures not carved by Bill Lemon. Based on a suit of armor in the Wallace Collection museum, London. Reissued as **474, 884**.

474 Red Knight: 1964-71, 1/8; **$20-30**
Also issued as **K-4, 884**. Box art by Künstler.

475 Gold Knight: 1965-69, 1/8; **$250-300**
Also issued as **K-5, 885**. Box art by Künstler.

825 King Arthur: 1968-69, 1/8; **$50-60**
Camelot set with new base and round shield. Gold plastic. Same base reading "Camelot" used in all three kits, but plastic color is randomly gray or green. This set was tied to the 1967 movie *Camelot*. Also issued as **Silver Knight K-1, 471, 881**. Box art by Schaare.

826 Sir Galahad: 1968-69, 1/8; **$50-60**
Camelot set with new base and diamond shaped shield. Silver plastic. Also issued as of **Blue Knight K-2, 472, 882**. Box art by Schaare.

827 Sir Lancelot: 1968-69, 1/8; **$50-60**
Camelot set with new base and arrowhead shaped shield. Blue plastic. Also issued as **Black Knight K-3, 473, 883**. Box art by Schaare.

881 Sir Galahad England: 1973-74, 1/8; **$50-60**
Knights in Shining Armour set. Metal plated. All models in this set have their new names engraved on the original bases. Also issued as **Silver Knight K-1, 471, 825**.

882 Sir Kay England: 1973-74, 1/8; **$50-60**
Knights in Shining Armour set. Metal plated. Also issued as **Blue Knight K-2, 472, 826**.

883 Sir Lancelot England: 1973-74, 1/8; **$50-60**
Knights in Shining Armour set. Metal plated. Also issued as **Black Knight K-3, 473, 827**.

884 Sir Percival England: 1973-74, 1/8; **$50-60**
Knights in Shining Armour set. Metal plated. Also issued as **Red Knight K-4, 474**.

885 King Arthur England: 1973-74, 1/8; **$180-200**
Knights in Shining Armour set. Metal plated. Also issued as **Gold Knight K-5, 475**.

886 Richard I England: 1973-74, 1/8
Knights in Shining Armour set. Metal plated. Also issued as **The Crusader K-7**. Crusader's spear was replaced with sword. No examples of this kit are known to exist; thus it may not have been issued.

Other Fighting Men Figure Kits from 1950s

K-6 The Viking: 1959-60, 1/8 **110-130**
Ivory plastic. Released in conjunction with Kirk Douglas's movie *The Vikings*. Model is based on Douglas, although clothing does not match that worn in the movie. Pattern by Bill Lemon.

K-7 The Crusader 59, 1/8; **$90-100**
Light blue plastic. Decals for cross on chest and lion on shield. Pattern by Bill Lemon. Box art by Ed Marinelli. Reissued with a sword instead of original spear in Knights in Shining Armour set as **886 Richard I**. Reissued in the original light blue plastic, but with the new sword in 1975 as Young Model Builders Club kit **766**. In 1963 Marx issued a six-inch plastic toy figure based on this model.

K-8 Athos 59, 1/8; **$70-90**
Ivory plastic. Real feather for hat, metal rods for sword and scabbard. Pattern by Bill Lemon. Included in **398** gift set.

K-9 Porthos 59, 1/8; **$70-90**.
Ivory plastic. Real feather for hat, metal rods for sword and scabbard. Pattern by Bill Lemon. Included in **398** gift set.

K-10 Aramis 59, 1/8; **$70-90**
Ivory plastic. Real feather for hat, metal rods for sword and scabbard. Pattern by Bill Lemon. Included in **398** gift set, also issued as **410 D'Artagnan**.

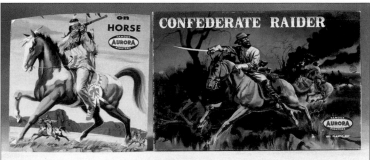

208 Two American Fighters: 1958-59, 1/8; **$350-400**
Gift set with **411 Infantryman, 412 Marine**. Includes four bottles of paint, brush, and cement.

216 Roman Gladiators 59, 1/8
Gift set includes **405, 406**. This gift set appears in Aurora's 1959 catalog, but evidently was not issued.

401 Apache Warrior: 1959-61, 1/8; **$300-350**
Ivory and brown plastic. Decals. Mounted on Black Fury **400** horse. Bow over shoulder, feathers in horse's mane, shooting rifle. Pattern by Bill Lemon. In England catalog as **408** in 73. Box art by Kotula.

402 Confederate Raider: 1959-61, 1/8; **$300-350**
Ivory and black or brown plastic. Mounted on Black Fury **400** horse. Raised sword in right hand. Pattern by Bill Lemon. Box art by Kotula.

398 Three Musketeers 59, 1/8; **$350-400**
Set includes **K-8, K-9, K-10**. Bases of the three models fit together in a semicircle. Box art is photo of built kits in front of street scene painted by Jim Cox.

405 Gladiator with Sword 59, 1/8; **$100-120**
Ivory plastic. Helmet lifts off. Rectangular base. Nameplate with scalloped corners reads, "Roman Gladiator 200 AD." Issued in **216**. Reissued as **405 Spartacus**. Box art by Kotula.

406 Gladiator with Trident 59, 1/8; **$100-120**
Ivory plastic, string net. Helmet lifts off. Rectangular base. Nameplate with scalloped corners reads, "Roman Gladiator 220 AD." Issued in **216**. Reissued as **406 Gladiator**. Both gladiator figures stand on rectangular bases with a decorative frieze around the sides. Neither figure carved by Bill Lemon. To save money Aurora hired a New Jersey sculptor to do them. He did not know how to break the figures down into parts as Lemon did; thus parting lines were cut at Ferriot Brothers. Box art by Kotula.

408 US Marshal 59, 1/8 ; **$100-120**
Ivory plastic. Stands with pistol in right hand over large marshal's badge on base. Pattern by Bill Lemon. Reissued as **408 Jesse James**.

409 U. S. Air Force Pilot: 1958-61, 1/8; **$90-110**
Gray-green plastic. "USAF" decal. Sun visor on helmet can be raised. Stands on base with half-globe and Air Force insignia disc. A better seller than the other US soldiers, perhaps because he has a more dynamic pose. Pattern by Bill Lemon. Reissued as **409 Steve Canyon**.

409 Steve Canyon 58, 1/8; **$200-220**
Gray-green plastic. Reissue of **409 Air Force Pilot** with new box art. Includes decal for helmet reading "CANYON." Inspired by the comic strip and short-lived TV series of the day. Kit issued so briefly that it did not appear in the catalog. Listed on dealers order sheet under number **404**.

410 U. S. Sailor: 1958-59, 1/8; **$25-35**
White plastic. "This was the quickest model I ever made," said Bill Lemon. Navy insignia disc on ship deck base. Box art by Ed Marinelli.

411 U. S. Army Infantryman 1957-59, 1/8; **$50-60**
Olive plastic. Rifle slung over shoulder. Decal sheet includes insignia for ranks from private to full sergeant. Stands on ground base with Army insignia disc. Issued with two box art versions: first shows soldier at attention (as in actual model), second art, by Jo Kotula, shows him in action pose exiting a helicopter. Pattern by Bill Lemon.

412 U. S. Marine: 1957-59, 1/8; **$50-60**
Dark blue plastic. Marine in dress uniform stands on base with Marine insignia disc. Includes sword and rifle, which can be assembled, held in right hand resting on shoulder or ground. Issued with two box art versions, both with the American flag in the background: first has sword on shoulder, second, by Jo Kotula, has rifle on shoulder. Pattern by Bill Lemon.

Guys and Gals of All Nations

All models are in ivory plastic. Inscribed patterns on parts represent fabric decorations.

209 Dutch Boy and Girl: 1958-59, 1/8; **$300-350**
Gift set includes glue and paint.

212 Indian Chief and Squaw: 1958-59, 1/8; **$300-350**
Gift set.

213 Chinese Mandarin and Girl: 1958-59, 1/8; **$300-350**
Gift set.

214 Scotch Lad and Lassie: 1958-59, 1/8; **$300-350**
Gift set.

215 Mexican Caballero and Senorita 59, 1/8
This gift set appears in the 59 catalog, but no copies of it have been found to date.

413 Dutch Boy: 1957-59, 1/8; **$30-40**
First of the Guys and Gals series to be issued. Nameplate reads "Netherlands." Name on box printed in either red or yellow. Pattern by Lemon.

414 Dutch Girl: 1957-59, 1/8; **$30-40**
First of the Guys and Gals series to be issued. Holds up empty left hand (contrary to box art which shows flowers). Nameplate reads "Netherlands." Pattern by Lemon.

415 Chinese Mandarin 1957-59, 1/8; **$30-40**
Nameplate reads "China." Pattern by Lemon.

416 Chinese Girl: 1957-59, 1/8; **$30-40**
Nameplate reads "China." Pattern by Lemon.

417 Indian Chief: 1957-59, 1/8; **$30-40**
Best seller of the Guys and Gals, now rare. Stands on rock base. Nameplate reads "American Indian." Pattern by Lemon.

418 Indian Squaw: 1957-59, 1/8; **$30-40**
Stands on ground base. Nameplate reads "American Indian." Pattern by Lemon. Box art by Ed Marinelli.

419 Scotch Lad: 1958-59, 1/8; **$30-40**
Nameplate reads "Scotland." Pattern by Lemon.

420 Scotch Lassie: 1958-59, 1/8; **$30-40**
Nameplate reads "Scotland." Pattern by Lemon.

421 Mexican Caballero 59, 1/8; **$130-150**
Last in the Guys and Gals series before it was discontinued, thus rare. Nameplate reads "Mexico." Pattern by Lemon.

422 Mexican Senorita 59, 1/8; **$130-150**
Nameplate reads "Mexico." Pattern by Lemon.

Wildlife

First issues are in 12x7" or 12x9" flat boxes and include paints and brush. In 1969 all but the Big Horn Sheep and Cougar were repackaged in 10x10" square boxes, and in 1972 in 8x8" square thin cardboard boxes.

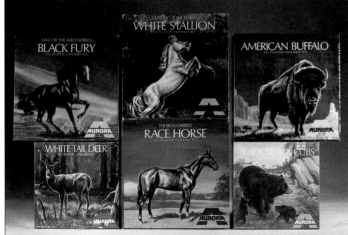

400 Black Fury: 1959-75, 1/8; **$30-70**
Black plastic. Unlike later Wildlife models, this has no base. Same horse used with Gold Knight **K-5**, Apache Warrior **401**, Confederate Raider **402**. First Wildlife Series kit and a very popular model, although sculptor Bill Lemon considered it one of his worst efforts. Contains only twelve parts, four of which are the horseshoes! Horse can be reared up to rest on hind legs and tail. First box art by Anthony Rudisill (although he signed it "Monaghan"). Second box art by Schaare.

401 White Stallion 1968-75, 1/12; **$50-60**
White plastic. Used in **808 Lone Ranger** and in **801 Zorro**. Creator Bill Lemon regarded this horse much more highly than his Black Fury. Box art by Schaare. Reissued as **446**.

402 American Buffalo: 1965-75, 1/16; **$45-55**
Brown plastic. Base has prairie dogs. Nameplate reads, "American Bison." Reissued as **445**. Pattern by Ehling. Box art by Steel.

403 White Tailed Deer: 1962-75, 1/8; **$40-50**
Brown-orange plastic. A ten-point buck. Squirrel on fallen tree trunk. Base reads, "White Tailed Deer." Pattern by Lemon. Box art by Steel.

404 Thoroughbred Race Horse: 1965-75, 1/12; **$40-50**
Dark tan plastic. Nameplate reads, "Thoroughbred." The prototype pattern's nameplate read "Yearling"—thus the kit appears in the catalogs as "Yearling Horse." Pattern by Ehling. Box art by Steel.

407 Black Bear & Cubs: 1962-75, 1/12; **$60-80**.
Black plastic. Mother and two cubs play by tree stump. Base reads, "Black Bear." First box art has hunter escaping by climbing rock. Hunter is deleted from box art in 1970s issues. Pattern by Lemon.

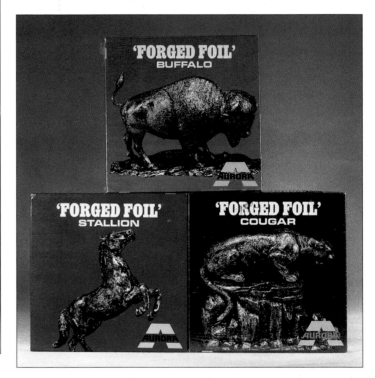

445 Forged Foil Buffalo 69, 1/16; **$50-60**
Black, brown plastic. Reissue of **402** with glue, aluminum foil, and paint. Added rectangular base below original base.

446 Forged Foil Stallion 69, 1/12; **$60-70**
Black, brown plastic. Revised issue of **401**. Added rectangular sub-base.

447 Forged Foil Cougar 69, 1/8; **$50-60**
Black, brown plastic. Revised issue of **453** without fawn. Added rectangular sub-base.

453 Cougar: 1963-66, 1/8; **$120-140**
Brown and gray plastic. Crouching cougar on rock threatens fawn. Two-part base reads, "Cougar." Pattern by Lemon. Reissued as **447**.

454 Big Horn Sheep: 1963-66, 1/12; **$110-130**
Brown and gray plastic. Mountain sheep poised to ram snarling lynx. Two-part base reads, "Big Horn Sheep." Pattern by Lemon.

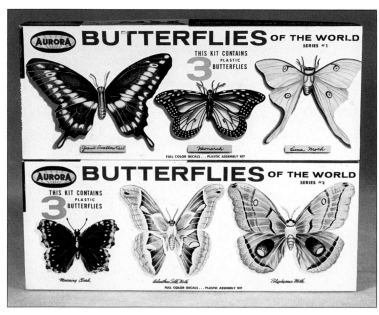

561 Butterflies of the World, Series 1: 1960-61, 1/1; **$30-40**
Beige plastic. Monarch, Luna Moth, Giant Swallow Tail. Decals for wing parts.

562 Butterflies of the World, Series 2 61, 1/1; **$40-50**
Beige plastic. Morning Cloak, Atlantic Silk Moth, Polyphemus Moth. Decals for wing parts.

Monsters

There are five Aurora monster series:

Movie Monsters, 1962-68. Thirteen models, most in 1/8 scale; In addition, **Gigantic Frankenstein,** the **Customizing** kits, **Munsters,** and **Addams Family** kits may be considered part of this series.

Glow Monsters, 1969-75. Twelve of the Movie Monsters reissued with glow parts. The first six were issued in the original flat boxes with "Frightening Lightning Strikes" labels added. The second six, as well as the original six, were issued in square boxes in 1970.

Monstermobiles, 1965-67. Six monsters in hot rod cars.

Monster Scenes, 1971. Eleven snap together kits issued as part of a diorama scene in 1/13 scale. Withdrawn due to public outcry. The final three kits in this series were released only in Canada.

Monsters of the Movies, 1975-77. Eight kits in 1/12 scale. Snap together construction and glow parts.

Movie Monsters

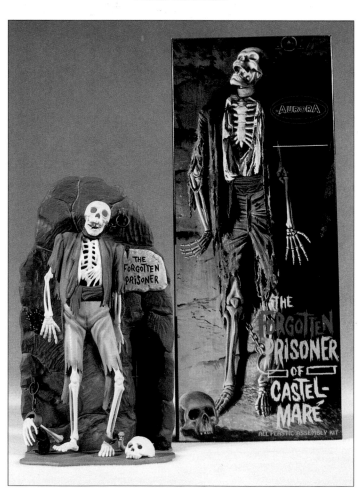

422 Forgotten Prisoner of Castel-Mare: 1967-68, 1/8; **$200-250**
Mustard plastic. Includes skull, snake, 2 rats. Stone on wall reads "The Forgotten Prisoner." Kit promoted by James Warren's *Famous Monsters* magazine company. Reissued as **453**. Pattern by Ehling. Box art by Künstler. Reissued by Cinemodels.

424 Dracula 1963-68, 1/8 ; **$200-250**
Black plastic. Base reads "Dracula." Based on Bela Lugosi in 1931 movie. Pattern by Lemon. Box art by Bama. Reissued as **454**. Reissued by Monogram and CineModels.

423 Frankenstein 1963-68, 1/8; **$200-250**
Gray plastic. Tombstone reads "Frankenstein." Based on Boris Karloff in 1931 movie. This model was manufactured from two duplicate molds; the base produced from one of the molds has both the Aurora and Universal names, while the other has only the Aurora trademark. Pattern by Lemon. Box art by Bama. Reissued as **449**. Reissued by Monogram.

425 Wolf Man 1963-68, 1/8; **$250-300**
Dark gray plastic. Rock base reads "The Wolf Man." Based on Lon Cheney, Jr. in 1941 movie. Pattern by Bill Lemon. Box art by Bama. Reissued as **450**. Reissued by Monogram.

426 Creature: 1963-68, 1/8; **$250-300**
Metallic green plastic. Base reads "The Creature." Original 1954 movie *Creature from the Black Lagoon* had no teeth, but Aurora's model has teeth. Pattern by Lemon. Box art by Bama. Reissued as **483**. Reissued by Monogram.

428 Phantom of the Opera 1964-68, 1/8; **$200-250**
Black plastic. Based on Lon Chaney's 1925 silent movie. Only monster kit to include a "victim" figure. Base reads, "The Phantom of the Opera." Pattern by Bill Lemon. Box art by Bama. Reissued as **451**. Reissued by Monogram and CineModels.

427 Mummy: 1963-68, 1/8; **$175-200**
Gray plastic. Temple ruins base reads, "The Mummy." Based on Boris Karloff as "Imhotep" in 1932 movie. Pattern by Bill Lemon. Box art by Bama. Reissued as **452**. Reissued by Monogram.

460 Dr. Jekyll as Mr. Hyde: 1965-68, 1/8; **$300-350**
White plastic. Nameplate reads, "Dr. Jekyll as Mr. Hyde." Reissued as **482**. Pattern by Ehling. Box art by Bama. Mold evidently destroyed.

461 Hunchback of Notre Dame: 1964-68, 1/8; **$200-250**
Tan plastic. Torture wheel base reads, "The Hunchback of Notre
Dame." Model based on Lon Chaney in 1923 silent movie. Pattern
by Bill Lemon. First box art by James Bama based on Anthony Quinn
in 1956 remake of the movie, but in 1966 the box art face was changed
by another artist. First version with Quinn's face is more valuable.
Reissued as **481**. Mold evidently destroyed.

469 Godzilla 1964-68, 1/200; **$250-350**
Fuchsia or purple plastic! Smashed Tokyo base nameplate reads,
"Godzilla." Model closely resembles Gozilla in the 1962 sequel *King
Kong vs. Godzilla*. Pattern by Meyers. Box art by Bama. Reissued as
466. Reissued by Monogram.

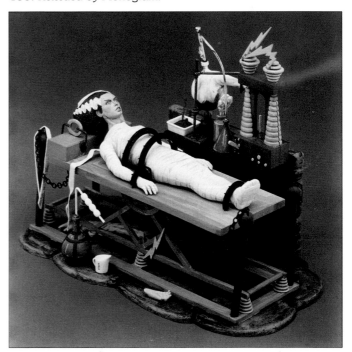

468 King Kong: 1964-68, 1/30; **$250-350**
Black and beige plastic. Carrying 1932 movie actress Fay Wray across
jungle island base that reads, "King Kong." Reissued as **465**. Pattern
by Ehling. Box art by Bama. Reissued by Monogram.

482 Bride of Frankenstein 1965-67, 1/11; **$450-500**
Gray plastic. *Poorly functioning mold limited production; never reissued. Thus today's high price.* Based on Elsa Lanchester in 1935
movie. Pattern by Meyers. Box art by Bama.

464 Customizing Monster Kit #2 64; **$100-120**
Light gray. Large vulture model, mad dog, skulls, rats, bats. Box art by Bama.

470 Gigantic Frankenstein 1964-65, 1/4; **$1200-1500**
Gray, bronze plastic. Includes three bottles of paint and brush. A toy like model with moving arms. Chain around monster's neck drags a piece of tombstone. An expensive $4.98 kit that sold poorly. "Big Frankie" stands almost two feet tall. Pattern by Ehling. Box art by Bama.

483 The Witch: 1965-67, 1/12; **$250-300**
Dark gray plastic. Not based on any movie. Reissued as **470**. Box art by Bama.

463 Customizing Monster Kit #1 64; **$100-100**
Light gray. Lizard from Creature kit, skull and rats from Wolf Man kit, bats from Dracula kit. Original Jake Smith and Willie Jones tombstones, giant spider and web, assorted bones, and rotten hands to replace regular hands of monsters. Box art by Bama.

804 The Munsters 65, 1/16; **$500-550**
Light gray and black plastic. Peel-and-stick papers for TV screen and "Home Sweet Home" painting over fireplace. Based on TV show. Living room scene with four characters. Box art by Bama.

805 Addams Family Haunted House: 1965-66, 1/64; **$400-500**
Light gray plastic. Haunted House with four moving ghosts activated by a lever running under the house. Cardboard pictures of family members can be cut out and pasted in windows not inhabited by ghosts. Box art by Bama.

Glow Monsters

Values given below for the first six glow monsters should be quadrupled for the 1969 long flat box "Frightening Lightning" issues. Canadian issues used wider boxes with colored panels on each side.

Later editions were packaged in square boxes. The kits in heavy cardboard square boxes with a 1969 copyright are valued more highly than those in light cardboard boxes with a 1972 copyright. To fit the square box format, some of James Bama's original long box art was widened by adding panels to each side, and then the background was overpainted to blend the added panels with the center section. Harry Schaare created new box art for some of the square boxes.

449 Frankenstein (GL): 1969-75, 1/8; **$140-160**
Black plastic. Added glow knot for rope belt. Reissue of **423**.

450 Wolf Man (GL): 1969-75, 1/8; **$200-240**
Dark gray plastic. Rarely purple plastic. Box art work of the Wolf Man changed from crouching to upright posture to conform with the pose of the model. Reissue of **425**. New box art by Schaare.

451 Phantom of the Opera (GL): 1969-75, 1/8; **$140-160**
Black plastic. Later issues gray plastic. Box art image reversed on square-box issue. Reissue of **428**.

452 Mummy (GL): 1969-75, 1/8; **$100-140**
Gray plastic. Rarely cream or black plastic. Reissue of **427**. Schaare copied Bama's original art to create a new "glow" version.

453 Forgotten Prisoner of Castel-Mare (GL): 1969-75, 1/8; **$200-250**
Light gray in Frightening Lightning issue. Dark gray plastic in square box issue. Reissue of **422**.

454 Dracula (GL): 1969-75, 1/8; **$140-160**
Black plastic. Reissue of **424**.

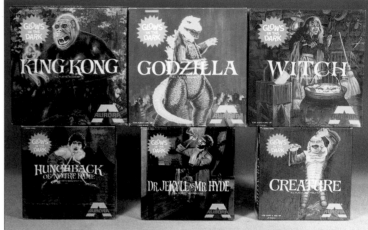

465 King Kong (GL): 1970-75, 1/30; **$160-180**.
Dark brown, light green plastic. Reissue of **468**. Schaare copied Bama's original art to create a new "glow" version.

466 Godzilla (GL): 1970-75, 1/200; **$160-180**
Metallic green plastic. Reissue of **469**.

470 Witch (GL): 1970-75, 1/12; **$120-140**
Black. Reissue of **483**. Schaare copied Bama's original art to create a new "glow" version.

481 Hunchback of Notre Dame (GL): 1970-75, 1/8; **$120-140**
Brown plastic. Reissue of **461**.

482 Dr. Jekyll as Mr. Hyde (GL): 1970-75, 1/8; **$180-220**
Orange-brown plastic. Includes spider that is not in original issue. Reissue of **460**. This is the final issue of this kit by any company. The mold evidently has been destroyed.

483 Creature from the Black Lagoon (GL): 1970-75, 1/8; **$180-200**
Metallic green plastic. Reissue of **426**.

458 Wolf Man's Wagon: 1965-67, 1/12; **$300-350**
Dark gray plastic. Pattern by Ehling. Box art by Bama.

466 Dracula's Dragster: 1965-67, 1/12; **$300-350**.
Black plastic. Box art by Bama.

459 Mummy's Chariot: 1965-67,
1/12; **$300-350**
Gray plastic. Box art by Bama.

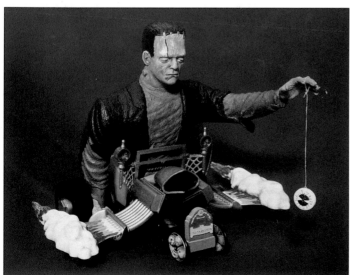

465 Frankenstein's Flivver: 1965-67, 1/12; **$300-350**
Gray plastic. Pattern by Ehling. Box art by Bama.

484 King Kong's Thronester: 1965-67, 1/30; **$400-500**
Black and yellow plastic. Not released until 1966. Larger kit than the earlier Monstermobiles. Pattern by Ehling. Box art by Vic Prezio.

485 Godzilla's Go Cart: 1965-67, 1/200; **$600-700**
Green and yellow plastic. Not released until 1966. Larger kit than earlier Monstermobiles. Box art by Vic Prezio.

Monster Scenes
Each figure kit has two sets of arms and legs. Snap-together assembly. Bases are small and basic, unlike traditional Aurora bases.

631 Dr. Deadly: 1971, 1/13 **100-110**
Gray plastic. Hunchbacked figure in apron with three sets of arms.

632 The Victim: 1971, 1/13 **100-110**
Ivory plastic. Issued in Canada as **Dr. Deadly's Daughter**. Pattern by Lemon. Box art by Schaare.

633 Frankenstein: 1971, 1/13; **$100-110**
Gray and glow plastic. Toylike, broad-shouldered figure. One set of arms allow him to carry the Victim. Box art by Schaare.

634 Gruesome Goodies: 1971, 1/13; **$100-120**
Brown and clear plastic. Dungeon laboratory equipment, electric generator, two tables, test tubes, skull, rat, and vampire rabbit.

635 Pain Parlor: 1971, 1/13; **$110-130**
Gray and glow plastic. Glow in the dark skeleton hangs on display, instrument wall panel, lab table for Frankenstein.

636 The Pendulum: 1971, 1/13; **$100-120**
Brown plastic. Swinging blade can be lowered.

637 Hanging Cage: 1971, 1/13; **$100-120**
Blue-gray plastic. Cage can be cranked up and down on a wooden gallows. Door opens, padlocks snap on and off. The Victim fits inside. Three part, snap-together base.

638 Vampirella: 1971, 1/13; **$150-200**
Tan plastic. A Warren Magazines character. Pattern by Lemon.

641 Dracula: Canada 1971, 1/12; **$400-420**
Black plastic. First run molded in US in glow plastic. One assembly option has candle in right hand, key in left. Turnstile under base allows any one of three creatures to be shown in the floor grating. Released only in Canada. Reissued in 75 as **656** in Monsters of the Movies set.

642 Dr. Jekyll & Mr. Hyde: Canada 1971, 1/12; **$400-420**
Fuchsia plastic. First run molded in US in glow plastic. Three heads: normal, mid-transformation, monster. Released only in Canada. Pattern by Lemon. Reissued in 75 as **654, 655** in Monsters of the Movies set.

643 Giant Insect: Canada 1971, 1/13; **$800-1000**
Green and clear. Released only in Canada. A scorpion's body and tail with a dragonfly's head and wings. A very rare kit.

Monsters of the Movies
Released in 1975, but first appeared in 1976 catalog. Each model has a movie maker's clack board which serves as a nameplate. The name squares are cut from the instruction sheets.

651 Frankenstein Monster: 1975-77, 1/12; **$150-170**
Black and glow plastic. Holds club in left hand. Design by Dave Cockrum. Reissued by Revell.

652 Wolf Man: 1975-77, 1/12; **$170-190**
Gray plastic. Lurks next to tree. Pattern by Ehling.

653 Creature: 1975-77, 1/12; **$250-300**
Metallic green and glow plastic. Arms rotate in shoulder sockets. Underwater scene with fish, anchor, and seaweed. Design by Cockrum. Pattern by Lemon.

654 Dr. Jekyll: 1975-77, 1/12; **$90-110**
White and glow plastic. Two heads and arms used from **642**. One normal head, one mid-transformation, both glow. Medical stand, beaker.

655 Mr. Hyde: 1975-77, 1/12; **$90-110**
Brown and glow plastic. Head and street lamp glow. Cobblestone base. Torso parts from **642**.

656 Dracula: 1975-77, 1/12 ; **$150-170**
Black and glow plastic. Body from **641**. Reissued by Revell.

657 Rodan: 1975-77, 1/300; **$200-250**
Brown and gray plastic. No glow parts. Licensed from Toho of Japan; from the 1957 movie and subsequent *Godzilla* movies. Design by Cockrum.

658 Ghidorah: 1975-77, 1/300; **$200-250**
Tan and orange plastic. No glow plastic. Licensed from Toho of Japan; from the 1965 movie. Design by Cockrum. Pattern by Meyers.

Comic Book Super Heroes

Superman was the first of eleven comic book heroes to be released in the 1960s. Eight of these first edition kits included name plates for their bases that would be deleted in the 1970s Comic Scenes reissues.

421 The Incredible Hulk: 1967-70, 1/10 180-220
Metallic green plastic. Marvel Comics character. Nameplate reads, "The Incredible Hulk." Reissued as **184**.

462 Superman: 1964-70, 1/8; **$140-160**
Light blue plastic. DC Comics character. First issue box has TV's George Reeves oil painting art work. Later box issue used the same line art style as the other comic book character kits and is more valuable. "S" engraved on chest and cape. Nameplate reads, "Superman." Reissued as **185**. Reissued by Monogram, MPC, Revell.

467 Batman: 1965-70, 1/8; **$270-300**
Blue plastic figure, brown base and tree. DC Comics character. Owl perched in tree, "Batman" carved on tree trunk, and bat insignia engraved on Batman's chest. Kit's part fit is poor. One of first figure patterns done by Larry Ehling. Box art by DC artist Carmine Infantino. Reissued as **187**. Reissued by MPC, Revell.

476 Captain America: 1967-70, 1/8; **$400-500**
Light blue plastic. Marvel Comics character. Striding over wall and splashing into puddle. Nameplate reads, "Captain America." Pattern by Meyers. Reissued as **192.**

477 Spider-man: 1967-70, 1/12; **$270-320**
Tan plastic. Spiderman on stairs throws web over sleeping villain. Marvel Comics character. Nameplate reads, "Spiderman." Reissued as **182**. Box art by Schaare.

478 Superboy: 1965-68, 1/8; **$180-200**
Blue plastic. "S" engraved on chest and cape. DC Comics character. Superboy and his dog confront a dragon in a cave. Nameplate reads, "Superboy and Krypto." Pattern by Meyers. Reissued as **186**.

479 Wonder Woman: 1967-68, 1/12; **$380-400**
Tan and light brown plastic. DC Comics character. Lassoing an octopus. Had a short life in catalog because, like other female figures, it sold poorly. Never reissued. Very scarce now. Nameplate reads, "Wonder Woman."

488 Robin: 1967-68, 1/8; **$140-160**
White plastic. DC Comics character. Decal for computer panel. Name plate reads: "Robin: The Boy Wonder." Face is flat and featureless, reportedly due to factory repairs on the mold to correct thin spots. Reissued as **193**. Reissued by Revell.

808 Lone Ranger: 1967-69, 1/10; **$200-220**
Medium blue and white plastic. On rearing horse, brandishing pistol in right hand—the same pose as in the title scene of the TV show. Horse is same mold as White Stallion **401**. No nameplate. Ranger figure by Ehling. Reissued as **188**. Box art by Schaare.

809 Tonto: 1967-69, 1/10; **$190-210**
Tan plastic. Pet eagle Taka perched on tree and rattlesnake coiled on ground. No nameplate. Pattern by Ehling. Reissued as **183**. Box art by Schaare.

820 Tarzan: 1968-69, 1/11; **$180-200**
Beige plastic. Stands over dead lion with right hand raised. Capitalized on NBC TV's *Tarzan* show staring Ron Ely. No nameplate. Pattern by Meyers. Reissued as **181**.

Comic Scenes

Ten of the 1960s comic book characters which had been dropped from the catalog were refurbished and reissued in 1974-75. Name plates which had gone with the original issues were deleted. Models were put in square boxes, along with instruction sheets in the form of a comic book.

181 Tarzan (CS): 1974-75, 1/11; **$70-90**
Tan plastic. Comic story by Mark Hanerfeld. Box and comic art by Neal Adams. Reissue of **820**.

182 Spider-man (CS): 1974-75, 1/12; **$140-160**
Red plastic. Comic story by Len Wein. Box and comic art by John Romita. Tin can and atomic pistol from first issue deleted. Reissue of **477**.

183 Tonto (CS): 1974-75, 1/10; **$50-70**
Brown plastic. Comic story by Marv Wolfman. Box and comic art by Gil Kane. Reissue of **809**.

184 Incredible Hulk (CS): 1974-75, 1/12; **$100-120**.
Light green plastic. Comic story by Len Wein. Box and comic art by Herb Trumpe. Reissue of **421**.

185 Superman (CS): 1974-75, 1/8; **$70-90**
Light blue plastic. Received new head parts, "S" removed from chest/cape and replaced with stick-ons. Comic story by Marv Wolfman. Box and comic art by Curt Swan. Reissue of **462**. Also reissued by Monogram and MPC (with new head parts).

186 Superboy (CS): 1974-75, 1/8; **$80-100**
Light blue plastic. Engraving removed and replaced by new stick-on "S" insignias. Comic story by Marv Wolfman. Box and comic art by Dave Cockrum. Reissue of **478**.

187 Batman (CS): 1974-75, 1/8; **$70-90**
Gray and brown plastic. Owl and "Batman" removed from tree, bat symbol removed from chest and replaced by stick-on label. Comic story by Len Wein. Box and comic art by Dick Giordano. Reissue of **467**. Also reissued by MPC with new head parts with sharper features.

188 Lone Ranger: 1974-75, 1/12; **$70-90**
Lighter blue than the original issue and white plastic. Comic story by Marv Wolfman. Box and comic art by Gil Kane. Reissue of **808**.

192 Captain America: 1975, 1/12; **$140-160**
Blue plastic. Released only in second year of series. Comic story by Len Wein. Box and comic art by John Romita. Water splash parts deleted from first issue. Reissue of **476**.

193 Robin 75, 1/8; **$100-120**
Ivory plastic. "R" logo removed from chest and replaced by stick-on. New head parts. Comic story by Marv Wolfman and Felton Marcus. Box and comic art by Dick Giordano. Reissue of **488**.

The Whoozis

Caricatures of real people. All models are in beige plastic. Paper stick-on strips with clever sayings fit on bases.

201 Susie Whoozis: 1967, 1/10; **$45-55**
"So what if I ain't smart, I'm lovely." Standing teenage girl.

202 Esmerelda Whoozis: 1967, 1/10; **$45-55**
"First mistake of the day—I got out of bed." Mom in bathrobe. Pattern by Ehling.

203 Denty Whoozis: 1967, 1/10; **$45-55**
"Caution: Engage brain before shifting mouth into gear." Talking girl seated on stool.

204 Alfalfa Whoozis: 1967, 1/10; **$45-55**
"I know that homework never killed anyone, but I'm not taking any chances." Teenage boy at desk listens to radio and ignores books.

205 Kitty Whoozis: 1967, 1/10; **$45-55**
"Don't Disturb While Doing Homework." Teenage girl sits on school books while applying makeup.

206 Snuffy Whoozis: 1967, 1/10; **$45-55**
"Hard work never killed anybody, but I'm not taking any chances." Dad reclines in a hammock.

Other Figure Kits from the 1960s

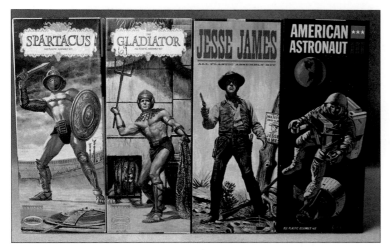

405 Spartacus: 1965-67, 1/8; **$100-120**
Tan plastic. New rectangular nameplate without scalloped corners reads "Roman Gladiator 200 AD." Reissue of **405 Gladiator with Sword** on a new arena diorama base. Box art by Künstler.

406 Gladiator: 1965-67, 1/8; **$70-90**
Tan plastic, string net. New rectangular nameplate without scalloped corners reads "Roman Gladiator 220 AD." Reissue of **406 Gladiator with Trident** with new diorama base. Box art by Künstler.

408 Jesse James: 1966-67, 1/8; **$120-140**
Tan plastic. Reissue of **408 US Marshal** with star on chest removed. Has new hat and new diorama base with cactus, snake, rifle, and stump with cardboard poster reading, "Wanted, Jesse James." Box art by Künstler.

409 American Astronaut: 1967-71, 1/12; **$70-80**
White and clear plastic. Figure has Astronaut Maneuvering Unit on his back. Base and background show Gemini capsule, Earth, and Moon. Box art by Schaare.

410 D'Artagnan: 1966-67, 1/8; **$50-70**
Tan plastic. Metal rods for sword and scabbard. Reissue of **K-10 Aramis** with new street scene base, new nameplate, and right arm lowered so that sword points straight ahead. Box art by Künstler.

411 Napoleon Solo: 1966-68, 1/12; **$200-220**
Tan plastic. Based on *The Man from U.N.C.L.E.* TV show which ran from 1964-68. Solo is climbing wall which interlocks with **412**. Nameplate reads, "The Man from U.N.C.L.E. Napoleon Solo." Box art by Künstler.

412 Ilya Kuryakin: 1966-68, 1/12; **$200-220**
Tan plastic. Kuryakin figure crouches behind a lamp next to wall which interlocks with **411**. Nameplate reads, "The Man from U.N.C.L.E. Ilya Kuryakin." Box art by Künstler.

413 Green Beret: 1967-69, 1/11; **$50-70**
Green plastic. Holds rifle in left hand, throws grenade with left. Jungle base nameplate reads "Green Beret, Special Service Forces." Pattern by Ehling. Box art by Künstler.

414 James Bond 007: 1967-68, 1/8; **$120-140**
Gray plastic. From Sean Connery of the *Goldfinger* movie. Holds pistol and crouches by wall. Nameplate reads "James Bond 007." Box art by Künstler.

415 Odd Job: 1967-68, 1/8; **$120-140**
Tan plastic. Villain in *Goldfinger*. Set to throw his hat. Nameplate reads "Odd Job." Pattern by Meyers. Box art by Künstler.

416 The Penguin: 1968-70, 1/12; **$350-400**
Black plastic. Decal sheet. From the *Batman* TV show. Holds umbrella in each hand while standing on a packing crate labeled "Penguin Umbrellas." Bat-a-rang lodged in crate.

451 The Frog: 1966-68, 1/1; **$140-160**
Green plastic. A "Castle Creature" inspired by movie producer Will-

iam Castle. Frog sits on lily pad next to sign reading, "Kiss me and you'll live forever—you'll be a frog, but you'll live forever." Pattern by Ehling.

452 The Vampire: 1966-68, 1/12; **$140-160**
Tan plastic. "Castle Creature." Stands next to cracked mirror and sign reading, "I like to go out at night—way out."

481 Hercules and the Lion: 1966-67, 1/8; **$275-300**
Tan, brown plastic. Pattern made for Tarzan fighting lion with pet ape looking on. When a license for Tarzan could not be secured, the ape was removed and a new Hercules head substituted. Nameplate reads, "Hercules and the Lion." Box art by Künstler.

455 Mad Barber Canada 66, 1/7; **$1000-1200**
Tan plastic. The "Mad Professionals" series was developed in the US by HMS, but was never issued in the US. Issued by Aurora Canada. They were announced in a Netherlands catalog in 1972, but apparently not issued. Only a very few kits have surfaced in the collectors' market

456 Mad Doctor: Canada 1966, 1/7; **$1000-1200**
Tan plastic. "Mad Professionals."

457 Mad Dentist: Canada 1966, 1/7; **$1000-1200**
Tan plastic. "Mad Professionals."

800 La Guillotine: 1964-68, 1/15; **$120-200**
Tan plastic. First in a projected "Chamber of Horrors" series from Madame Toussaud's Wax Museum. Plank can be raised and lowered to put victim under the blade. Blade can be raised and released. Nameplate reads "Guillotine." Toussaud's name removed from package in later issues. Pattern by Ehling. Box art by Bama.

463 Blackbeard: 1966-67, 1/10; **$230-250**
Tan plastic. "Blackbeard" decal. The rarer of the two "Bloodthirsty Pirates." Base is ship's deck with wheel and treasure chest. The most dynamically posed of all Aurora figures! Pattern by Ehling.

464 Captain Kidd: 1966-67, 1/10; **$50-70**
Beige plastic. "Bloodthirsty Pirates." Base reads "Captain Kidd." Figure holds sword in right hand, pistol in left. Includes treasure chest, money bags, barrel, skull, scorpion, lantern. Pattern by Ehling.

480 Captain Action: 1967-70, 1/6; **$350-380**
Dark metallic blue plastic. "CA" decal for chest. Base reads "Captain Action." Licensed character from Ideal Toys, which produced an action figure nearly identical to Aurora's model kit. Pistol in right hand, knife in left. At one-foot tall, he is larger than all figure models, except Gigantic Frankenstein. Pattern by Ehling.

801 Walt Disney's Zorro: 1966-68, 1/12; **$250-300**
Black plastic. Issued when Disney's 1950s series was being rerun on TV. Base with nameplate reads, "Zorro." On same horse used in **808, 401.** Pattern by Lemon. Box art by Künstler. Issued in Netherlands in 1974-75.

802 Alfred E. Neuman: 1965, 1/12; **$70-80**
Beige plastic. Based on *Mad* magazine character. Comes with four sets of snap-on arms and four sayings on cardboard which slip into a sign: "Love Thy Neighbor, Listen to the Voice of Experience, Honesty is the Best Policy, Down with School, Homework, etc." Pattern by Ehling.

806 Nutty Nose-Nipper Machine: 1965, 1/8; **$180-200**
Tan and beige plastic. Red plastic nose. Metal spring and squeeze-box noise maker. Torture device to pull nose of victim by turning Rube Goldberg crank to tighten noose around victim's nose. Nameplate reads "Probascus Tortus."

807 Wacky Back-Whacker Machine: 1965, 1/8; **$130-150**
Tan and beige plastic. Metal spring, string, squeeze-box noise maker. Nameplate reads: "Posterior Paddle." Turning a crank causes gears to turn and paddle to whack victim's rump.

851 John F. Kennedy: 1965-68, 1/8; **$140-160**
Black plastic clothing, white body parts, background and base. Cloth American flag. Stick-on painting of PT-109 (copied from Revell's PT-109 box art!) Pattern by Meyers. Box art by Künstler.

852 George Washington: 1967, 1/12; **$120-140**
Gray and beige plastic. Model made in 1965 but not released until 1967. Inspired by Emanuel Leutze's "Washington Crossing the Delaware." Large background part has boat and ten figures. Washington and secondary soldier figure stand on foreground shore base part.

814 Doctor Dolittle—Pushmi-Pullyu: 1968-69, 1/15; **$70-80**
Gray plastic. Based on the 20th Century Fox Doctor Dolittle movie. Aurora's first snap-together kit. Figures stand on base reading "Doctor Dolittle and the Pushmi-Pullyu." Pattern by Meyers.

815 Good Ship Flounder from Doctor Dolittle: 1968-69, 1/72; **$80-100**.
Black and white plastic. Ship with small figures of Doctor Dolittle, Emma, Tommy, Matthew, monkey, pig, and seal.

818 Dick Tracy: 1968-69, 1/16; **$70-80**
Dark blue plastic. Based on the comic strip and TV show. Tracy climbs down fire escape with pistol in right hand. **819 Space Coupe** is a companion kit. Pattern by Meyers.

853 Historic Flag Raising at Iwo Jima: 1966-68, 1/15; **$150-170**
Dark olive green plastic. Two peal-and-stick paper flag halves. Six figures. Base reads, "United States Marines Raising Flag on Mount Suribachi, Iwo Jima, World War II February 23rd 1945." Pattern by Meyers. Box art by Künstler.

860 Willie Mays (GM): 1966, 1/10; **$175-200**
Light gray plastic. Decal sheet for uniform markings. Great Moments in Sports series. All the instruction sheets in this "Great Moments in Sports" series contain a biographical profile written by Fred Katz, of *Sport* magazine. Model of Mays catching fly ball over his shoulder at 450 ft. mark on fence. Nameplate reads, "Willie Mays, 1954 World Series, Polo Grounds." Box art by Künstler.

861 Dempsey-Firpo (GM): 1965-66, 1/14; **$80-100**
Tan, beige plastic. Great Moments in Sports. One corner of a boxing ring, two fighters, and a referee. Nameplate reads, "Dempsey-Firpo, September 14, 1923, New York, N. Y." Pattern by Meyers. Box art by Künstler.

862 Babe Ruth (GM): 1966, 1/14; **$240-260**
Tan and green plastic. Great Moments in Sports. Ruth, catcher, umpire. Cardboard backdrop of stadium crowd. Nameplate reads, "Babe Ruth, 60th home run of 1927, Yankee Stadium." Box art by Künstler.

863 Jimmy Brown (GM): 1966, 1/13; **$120-140**
Light gray and dark green plastic. Uniform numbers decal. Great Moments in Sports. Brown lunges past two defenders toward goal posts. Nameplate reads, "Jimmy Brown, Breaks NFL Ground Gaining Record—More Than 10,000 Yards." Box art by Künstler.

864 Johnny Unitas (GM): 1966, 1/13; **$120-140**
Tan, green plastic. Uniform numbers decals. Great Moments in Sports. Unitas passes, lineman blocks defensive player. Nameplate reads, "Johnny Unitas, breaks NFL passing record with 237 completions in 1963 season." Pattern by Lemon. Box art by Künstler.

865 Jerry West (GM): 1966, 1/12; **$120-140**.
Tan plastic. Uniform numbers decal. Great Moments in Sports. West shoots over defender's hands. Includes complete backboard and stand. Box art by Künstler.

Prehistoric Scenes

Bases interlock into one large diorama. Snap construction and moving parts. Monogram and Revell have reissued most of the animal kits.

729 Neanderthal Man: 1972-76, 1/13; **$70-90**
Light brown plastic. First box art shows Allosaurus head, which was deleted from second, larger box to avoid impression dinosaur was included. Holds boulder or spear. Two sets of arms and legs. Nameplate reads, "Neanderthal Man." Lemon did the pattern, but said of the cave people figures: "They were snap together, and as far as I was concerned they were lousy."

730 Cro-Magnon Man: 1972-76, 1/13; **$70-90**
Tan plastic. Two sets of arms and legs. Nameplate reads, "Cro-Magnon Man." Pattern by Lemon.

731 Cro-Magnon Woman: 1972-76, 1/13; **$70-90**
Tan plastic. Second edition in larger box. Two sets of arms and legs. Sculptor Bill Lemon was asked to create a second pattern because his first was too busty and had too short a skirt! Nameplate reads, "Cro-Magnon Woman."

732 The Cave: 1972-76, 1/13; **$50-70**
Tan plastic. First box art has Allosaurus outside cave, which is removed in second issue to avoid impression dinosaur was included. Nameplate reads, "Cave, Home of Early Man." Two Cave kits can be locked together to form a complete enclosure.

733 Saber Tooth Tiger (Smilodon): 1972-76, 1/13; **$90-100**
Orange plastic. Two sets of movable legs. First edition box art shows snake. Second edition comes in larger box with snake deleted and contains a second base part. Nameplate reads, "Sabertooth Tiger." Pattern by Meyers.

734 Flying Reptile (Pteranodon): 1972-76, 1/13; **$45-65**
Red-orange plastic. Head pivots on ball-in-socket joint, wing tips flap on hinges, extra "battle scarred wing." Fin on back serves as attachment for string for hanging. Nameplate reads, "Pteranodon." Pattern by Lemon. Reissued by Monogram, Revell.

735 Tar Pit: 1972-76, 1/13; **$120-140**
Orange plastic. Woolly Rhino stuck in tar and vulture in tree. Nameplate reads, "La Brea Tar Pit." Pattern by Meyers.

736 Allosaurus: 1972-76, 1/13; **$60-80**
Metallic green plastic. First edition box art shows Saber Tooth Tiger attacking Allosaurus and contains one base part. Second edition comes in a larger box, deletes tiger from art and includes a two-part base. Only model in series not sculpted by Lemon or Meyers. Nameplate reads, "Allosaurus." Reissued by Monogram, Revell.

738 Cave Bear: 1973-76, 1/13; **$70-90**
Dark brown plastic. Original pattern made in 1960s as Grizzly Bear for Wildlife series, but never produced. Nameplate reads, "Cave Bear." Pattern by Meyers. Monogram reportedly still retains this mold, but has not reissued the kit.

739 Giant Bird (Phororhacos): 1973-76, 1/13; **$70-90**
Metallic blue plastic. Includes hatching eggs, small winged lizard, three alternate legs, and four alternate feet. Nameplate reads "Phororhacos." Pattern by Lemon.

740 Jungle Swamp: 1973-76, 1/13; **$130-150**
Orange and light green plastic. Pond, cycad trees, archaeopteryx, two pteranodons, snake, winged lizard, small dinosaur, and Dawn Horse. No name plate.

741 Three Horned Dinosaur: 1973-76, 1/13; **$80-90**
Metallic gray plastic. Same body parts as **742**. Nameplate reads "Triceratops." Pattern by Lemon. Reissued by Monogram, Revell.

742 Spiked Dinosaur: 1973-76, 1/13; **$80-90**
Tan and green plastic. Same body parts as **741**. Nameplate reads, "Styrachosaurus." Pattern by Lemon. Reissued by Monogram, Revell.

743 Giant Woolly Mammoth: 1973-76, 1/13; **$100-120**
Brown and white plastic. First of the Prehistoric figures carved by Lemon. Nameplate reads, "Woolly Mammoth." Reissued by Monogram, Revell.

744 Armored Dinosaur (Ankylosaurus): 1974-76, 1/13; **$130-150**
Orange plastic. Nameplate reads, "Ankylosaurus." Pattern by Lemon. Reissued by Monogram, Revell.

745 Sailback Reptile (Dimetrodon): 1974-76, 1/13; **$130-150**
Bronze and green plastic. Tongue, head, legs, tail move. Includes salamander, dragonfly, spider. Nameplate reads, "Dimetrodon." Pattern by Lemon. Reissued by Monogram, Revell.

746 Tyrannosaurus Rex: 1974-76, 1/13; **$300-350**
Orange plastic. Eyes, claws, and teeth in glow plastic. A large model—almost a yard long from nose to tip of tail—that was issued just at the height of the oil shortage! Nameplate reads, "Tyrannosaurus Rex." Cardboard backdrop. Design by Dave Cockrum. Pattern by Lemon. Reissued by Monogram, Revell.

Aurora/ESCI Figure Kits

Manufactured in Italy by ESCI, packaged in the United States by Aurora.

6210 Marine Assault Group: 1977, 1/72; **$5-6**
6211 German Infantry: 1977, 1/72; **$5-6**
6212 Nebelwerfer Battery: 1977, 1/72; **$5-6**
6224 British 8th Army Infantry: 1977, 1/72; **$5-6**
6230 Russian Red Guards: 1977, 1/72; **$5-6**
6301 Nebelwerfer Battery: 1977, 1/35; **$8-10**
6302 British Red Devil Paratroopers: 1977, 1/35; **$5-6**
6303 German Engineers: 1977, 1/35; **$5-6**
6304 German Green Devils: 1977, 1/35; **$5-6**
6305 European Partisans: 1977, 1/35; **$5-6**
6307 German Command Post: 1977, 1/35; **$5-6**
6308 Storm Troopers: 1977, 1/35; **$5-6**

Miscellaneous

451 Totem Craft: 1957-59, 1/10; **$60-80**
Four 9", one-piece, pre-shaped white styrofoam totem poles. Kits came with either a cardboard tray of water colors or four bottles of Aurora enamel paint. Manufactured for Aurora by Mattie Sullivan.

452 Alaskan Totem Craft: 1959, 1/10; **$80-100**
Three 9" white styrofoam totem poles. Paint.

551 Winchester 94 (Be): 1959-61, 1/3; **$50-60**
Ex-Best. Light brown, dark brown, and black plastic.

552 Mannlicker Rifle (Be): 1959-61, 1/3; **$50-60**
Ex-Best. Light brown, dark brown, and black plastic.

553 Savage 99 (Be): 1959-61, 1/3; **$50-60**
Ex-Best. Light brown, dark brown, and black plastic.

821 Tarzan King of the Jungle: 1968-69; **$180-200**
Beige plastic. Small plastic toy figures on a card. Tarzan, Cheetah and seven animals.

822 Oz-Kins: 1968-69; **$140-160**
Beige plastic. Eleven 2.5" plastic figures on a card. Dorothy, Toto, and the others. Based on Saturday morning ABC-TV show. Burry Biscuit Co. of Brooklyn offered these figures with four jars of Aurora paint and a brush in a cardboard mailer for $1 and a coupon. Patterns by Ehling.

823 Camelot: 1968-69; **$100-120**
White plastic. Ten 2" toy plastic knights and ladies on a card. Based on the movie *Camelot*.

824 NFL Miniatures: 1968; **$80-100**
Blue and white plastic. Thirty-four 2" figures. Coaches, officials, players. Includes jersey numbers decal sheet. Packaged in a box, not on a card. Box art by Schaare.

981 The Aquarium: 1955, 1/1; **$150-170**
Only Aurora kit designed by Gowland. 8.5" x 6" tank. Puerto Rico's most famous sculptor Alberto Vadi Rivera carved the pattern. The parts were manufactured at Gowland's plant in Caguas, Puerto Rico, but boxed by Aurora in the United States.

Snaparoos

This set of twelve very small snap-together kits was manufactured by R & L in Australia and packaged in the United States by Aurora. Marketed by Aurora's toy division.

9251 Jet Liners: 1975; **$8-10**.
Boeing 707, 727, Fokker F-27, Concorde

9252 Light Aircraft: 1975; **$8-10**
Spirit of St. Louis, Bell 204B, Beaver, Cessna 172

9253 Antarctic Explorers: 1975; **$8-10**
Dog sled, snow tractor, explorer and seal, explorer and penguin

9254 Explorers of Space: 1975; **$8-10**
Astronaut, Apollo LEM, radio telescope

9255 World of Ships: 1975; **$8-10**.
Queen Elizabeth, Canberra, Tugboat, Hovercraft, Mayflower

9256 Harbor Fleet: 1975; **$8-10**
Tanker, Freighter, Amphibious Duck, Police launch, Car ferry

9257 Old Time Cars: 1975; **$8-10**
Morris Oxford, 1904 Mercedes Racer, Model T Ford, 1926 Bugatti GP

9258 Old Timers: 1975; **$8-10**
1901 Oldsmobile, 1912 Packard, 1913 Sunbeam, 1927 Bentley

9259 Grand Prix Racers: 1975; **$8-10**
Brabham, Ferrari, Honda, BRM

9260 Heavy Movers: 1975; **$8-10**
Flat bed truck, Fork lift, Timber transport

9261 Big Hooks: 1975; **$8-10**
XKE Jaguar, Tow truck, Construction crane

9262 Historical Transportation: 1975; **$8-10**
Royal coach, North Star locomotive, San Francisco cable car, McLaren traction engine

HO Buildings

Aurora's first structure kits were designed for model railroads; later kits were made for slot car layouts. After 1962, the original HO railroad kits and Gas Station were manufactured by Tru-Scale. In 1972 Tyco reissued all but the Trees and Diner as pre-assembled models, and in 1974 Tyco added lights to the buildings.

651 Model Tree Kit: 1958-62, 1/87; **$50-70**
Brown trunk parts with green branch parts that stack up to form eight pine trees.

652 Colonial House: 1958-62, 1/87; **$45-55**
Gray and green plastic. Two-story house with sun deck over garage. Opening garage door. Tru-Scale **652**. Tyco **945**.

653 Ranch House: 1958-62, 1/87; **$45-55**
Gray and green plastic. One-story house with two-car garage. Opening garage door. Tru-Scale **653**. Tyco **944**.

654 School House: 1958-62, 1/87; **$60-70**
Maroon, dark gray, and clear plastic. "Aurtown School 1958" with playground equipment, flagpole, and paper flag. Tru-Scale **654**. Tyco "Factory" **940**.

655 Church: 1958-62, 1/87; **$60-70**
Light gray and dark gray plastic. Paper sheet for stained glass windows. Tru-Scale **655**. Tyco **941**.

656 Railroad Station: 1958-62, 1/87; **$60-70**
Gray, brown plastic. Tru-Scale **655**. Tyco **942**.

657 Joe's Diner: 1960-62, 1/87; **$60-70**
White and silver plastic. Converted trolley car. Named for Aurora founder Joe Giammarino. Tru-Scale **657**.

658 HO-Gas Station: 1961-62, 1/87; **$150-170**
"Model Motoring Service Center." White, red, gray, clear plastic. Garage bay doors open. First issued with "Aurora" sign decals; later "Texaco." Tru-Scale **658**. Tyco "Union 76" **943**.

1450 Start-Finish Pylons: 1963-73 MM 1/87; **$15-20**
White plastic. Metal springs. Green and checkered flag decals. Reissued in **1498**.

1451 Judges Stand: 1963-73 MM 1/87; **$40-60**
Gray plastic. Decals. Paper flags sheet. Reissued in **1498**.

1452 Grandstand: 1963-73 MM 1/87; **$40-60**
Brown and gray plastic. Decals. Reissued in **1499**.

1453 Double Station Pit Stop: 1963-73 MM 1/87; **$50-60**
Beige, gray plastic. Decals. Reissued in **1499**.

1456 Curved Bleachers: 1963-73 MM 1/87; **$40-60**
Dark brown plastic. Decals. Reissued in **1499**.

1498 Start-Finish Pylons & Judges Stand: 1974-77 MM 1/87; **$50-60**
Reissue of **1450, 1451**. Both kits in white plastic.

1499 Grandstand, Dual Pit Stop, Curved Bleachers: 1974-77 MM 1/87; **$70-80**
Reissue of **1452, 1453, 1456**. Grandstand in white and light green plastic. Dual Pit Stop in light brown and light orange plastic. Curved Bleachers in yellow plastic.

Monogram-Revell Reissues

Since 1978 Monogram has reissued a number of kits from Aurora tooling. Following the merger of Monogram and Revell into one corporation, some ex-Aurora models have appeared under the Revell trademark.

The kits below are listed by their Monogram or Revell numbers with the original Aurora kit numbers in parenthesis.

Aircraft

5203 (106) Fokker D-7: 1979-80, 95; **$10-12**
Reissued in 1995 in "Selected Subjects" series in the same 1979 box with a new copyright added. Also reissued as **74005**.

5204 (102) Sopwith Camel: 1979-80; **$10-12**
Also reissued as **74004**.

5205 (103) SE-5: 1979-80; **$10-12**
Fuselage extensively reworked to represent an SE-5, not the SE-5A. Also reissued as **74012**.

74004 (102) Sopwith Camel: 1992; **$10-12**
Reissue by Revell-Monogram of Germany under Monogram trademark. Also reissued as **5204**.

74005 (106) Fokker D-7: 1992; **$10-12**
Reissue by Revell-Monogram of Germany under Monogram trademark. Also reissued as **5203**.

74012 (103) SE-5A: 1992; **$10-12**
Reissue by Revell-Monogram of Germany under Monogram trademark. Also reissued as **5205**.

Ships

5412 (361) Boeing 747 Pan American 1978-79; **$15-20**

5413 (366) DC-10 American 1978-79; **$15-20**

5413 (366) DC-10 American: 1978; **$20-25**
Issued by Monogram/Nicomisa in Mexico.

5413A (366) DC-10 Mexicana: 1978; **$20-25**
Issued by Monogram/Necomisa in Mexico.

5414 (354) Boeing 727 TWA: 1978-79; **$15-20**

1514 (354) Boeing 727 TWA: 1978; **$20-25**
Issued by Monogram/Necomisa in Mexico.

1514A (354) Boeing 727 Mexicana: 78; **$20-25**
Issued by Monogram/Necomisa in Mexico.

5415 (359) Boeing 737 United: 1978-79; **$15-20**

3101 (711) Skipjack: 1979-81; **$15-20**

3102 (716) Wolfpack U-Boat: 1979-81, 95; **$15-20**
Reissued in 1995 in the "Selected Subjects" series in the same box as the 1979 issue with a new copyright added.

3103 (728) Japanese Submarine I-19: 1979-81; **$15-20**

(709) Graf Spee: 1979; **$40-50**
Issued in New Zealand without kit number and does not appear in US catalog.

(713) Yamato: 1979; **$40-50**
Issued in New Zealand without kit number and does not appear in US catalog.

3503 (700) Independence: 1978-86; **$20-25**

3504 (701) Forrestal: 1978-86; **$20-25**

3505 (702) Saratoga: 1978-86; **$20-25**

3700 (720) Enterprise: 1978-87; **$30-35**
Nuclear powered carrier. Also reissued as **75001**.

3707 (720) USS Enterprise: 2002; **$30-35**
Issued under the Revell trademark.

5071 (705) USS Missouri 2002; **$12-15**
Issued by Revell Germany under Revell trademark.

75001 (720) Enterprise: 1991; **$30-35**
Reissue by Revell-Monogram of Germany under Monogram trademark. Also reissued as **3700**.

5418 (395) A-7 Corsair: 1979-87; **$10-12**
Mold was extensively retooled.

5804 (368) F-111A: 1981-87, 91-93; **$15-17**
Mold was extensively retooled.

Tanks

6034 (317) Ground Attack: 1984; **$35-40**
Diorama kit with ex-Aurora Sherman Tank **317** and Monogram FW-190.

6035 (302) Tank Hunter: 1984; **$35-40**
Diorama kit with ex-Aurora Panther Tank **301** and Monogram P-51.

1/32 Cars

2000 (665) Mustang: 1980-82; **$20-25**
An air scoop was added to the hood to make it a 350 GT.

2001 (677) Mako Shark: 1980-82; **$15-20**

2002 (664) Pontiac GTO: 1980-83; **$15-20**

2003 (667) Plymouth Barracuda: 1980-82; **$15-20**

1/25 Cars

2243 (566) Jaguar XKE: 1979-80; **$15-20**

2244 (563) Ferrari Berlinetta GTO: 1979-80; **$15-20**
Also reissued as **77005**.

2245 (564) Maserati 3500 GT: 1979-80; **$15-20**
Also reissued as **77002**.

2246 (562) Aston Martin DB-4: 1979-80; **$15-20**
Also reissued as **77003**.

2907 (566) Jaguar XKE: 1990; **$10-12**

2954 (561) Porsche 904: 1992; **$10-12**

Revell-Monogram of Germany released the three cars below under Monogram trademark.

77002 (564) Maserati 3500 GT: 1992; **$15-17**
Also reissued as **2245**.

77003 (562) Aston Martin DB-4: 1992; **$15-17**
Also reissued as **2246**.

77005 (563) Ferrari Berlinetta GTO: 1992; **$15-17**
Also reissued as **2244**.

556 XKE Jaguar: 1998; **$10-15**

562 Aston Martin DB 4: 1997; **$10-15**

565 Ford GT: 9197; **$10-15**

Science Fiction

6011 (817) Flying Sub: 1979-80, 1995; **$20-25**
Reissued in 1995 in the "Selected Subjects" series in the same box as 1979 issue with a new copyright date.

6012 (813) U.F.O.: 1979-80, 1996, 2003; **$20-30**
Reissued in 1996 and 2003 in the "Selected Subjects" series in the same box as the 1979 issue with new copyright dates.

Dinosaurs

Monogram reissued two sets of dinosaurs, one in 1979-80 and a second in 1987. Different kit numbers were used for each issue. Kit numbers listed are for the 1979-80 issue, followed by the 1987 issue, and the original Aurora issue.

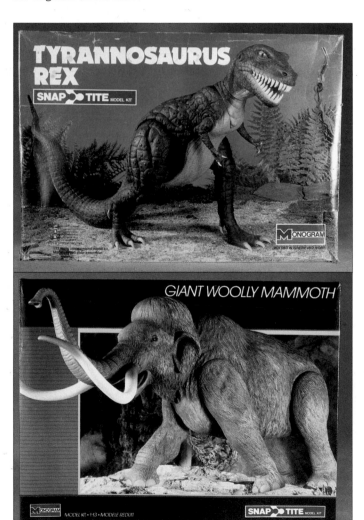

6040, 6074 (741) Three Horned Dinosaur: $20-25
6042, 6076 (742) Spiked Dinosaur: $20-25
6043, 6077 (746) Tyrannosaurus: $20-45
6044, 6078 (736) Allosaurus: $20-25

6045 (744) Armored Dinosaur: $30-40
The only dinosaur not reissued again by Monogram in 1987.

6046, 6080 (745) Dimetrodon: $20-25
6047, 6075 (743) Woolly Mammoth: $20-25
6079 (734) Flying Reptile: $20-25
The only dinosaur reissued in 1987 that had not been reissued in 1979-80.

In 1993 Revell of Germany issued all eight of the dinosaur kits under the Revell trademark. Five of these were also released by Revell of the United States during the summer of 1993 when the movie *Jurassic Park* premiered. Kit numbers listed are for the German issue, followed by the US issue, and the original Aurora issue. Prices for German issues are higher than US issues.

H6470, 6339 (746) Tyrannosaurus: 1992; **$30-50**
H6471, 6338 (741) Triceratops: 1992; **$15-30**
H6472 (742) Styrachosaurus: 1992; **$30-35**
H6473 (745) Dimetrodon: 1992; **$30-35**
H6474, 6336 (736) Allosaurus: 1992; **$15-30**
H6475, 6335 (734) Pteranodon: 1992; **$15-30**
H6476 (743) Mammoth: 1992; **$30-35**
H6477, 6337 (744) Ankylosaurus: 1992; **$15-30**

Figure Kits

6300 (469) Godzilla: 1978-79; **$50-60**
6301 (462) Superman: 1978-79; **$30-40**

In 1983 a set of four movie monsters was issued with glow parts.

6007 (423) Frankenstein: 1983; **$30-40**
6008 (424) Dracula: 1983; **$30-40**
6009 (425) Wolf Man: 1983; **$30-40**
6010 (427) Mummy: 1983; **$30-40**

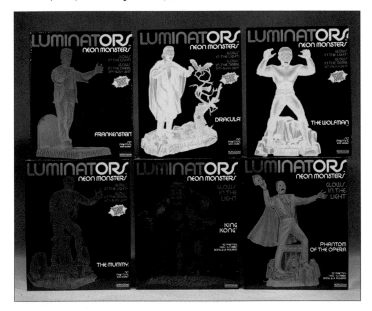

In 1991 a set of four monsters was issued in translucent "luminator" plastic. A black light could be ordered for glow-in-the-dark display of the built kits. In 1992, the Phantom and King Kong were added to this set.

1619 (423) Frankenstein: 1991; **$10-15**
1620 (424) Dracula: 1991; **$10-15**
1621 (425) Wolf Man: 1991; **$10-15**
1622 (427) Mummy: 1991; **$10-15**
1623 (468) King Kong: 1992; **$20-30**
When this kit reached the stores in the fall of 1992, it sold out within a few weeks. Because of mold defects, some small parts to the base are missing from this reissue.

Monogram reissued four "Special Value Pack" kits in 1994 for sale only at Wal-Mart during the Halloween season. They have no glow plastic parts; three bottles of Revell glow paint, a brush, and a tube of cement included.

6375 (423) Frankenstein 1994 **15-20**
6378 (427) Mummy: 1994; **$15-20**
6379 (425) Wolf Man: 1994; **$15-20**
6380 (424) Dracula: 1994; **$15-20**

6490 (426) Creature from the Black Lagoon: November 1994; **$20-25**
No glow parts.
6491 (428) Phantom of the Opera: November 1994; **$15-20**
No glow parts.

85-3633 (651) Frankenstein: 1999; **$20-25**
Molded from original Monsters of the Movies tooling. An exclusive at Toys "R" Us.

85-3634 (656) Dracula: 1999; **$20-25**
Molded from original Monsters of the Movies tooling. An exclusive at Toys "R" Us.

3635 (462) Superman: 1999; **$20-25**
3636 (467) Batman: 1999; **$20-25**
3637 (488) Robin: 1999; **$20-25**

3639 (802) Alfred E. Neuman: 2000; $20-25

Reissues by Other Companies

CineModels Reissues
In 1991 hobbyist Andrew Eisenberg formed CineModels to package and distribute monster models produced for him by Monogram. In the mid-1990s the Aurora trademark expired, and in 1999 CineModels, Inc. was granted the US trademark for the name Aurora inside the oval logo.

6008 (424) Dracula: 1991; **$20-25**
In the 1983 Monogram box with an added 1991 copyright date. No glow parts.

422 (422) The Forgotten Prisoner of Castle-Mare: 1993; **$25-35**
Facsimile of the original long box Aurora edition with added 1992 copyright.

428 Phantom of the Opera: 1994; **$25-35**
Facsimile of original long box Aurora edition with added 1993 copyright date.

Glencoe Reissues
In early 1996, Glencoe Models reisued some long-gone Aurora World War I biplane models using molds loaned by Monogram in exchange for removing all of the raised decal locater lines not previously removed by Aurora. A sheet of new decals helped bring them closer to modern modeling standards. Unlike Monogram's reissues, these kits include the mini-diorama ground base and ground crew figure. Another link with the past was new box art by John Amendola.

05114 (108) Nieuport 28C.1: 1996; **$10-12**
05115 (109) Pfalz D. III: 1996; **$10-12**
05118 (107) Spad XIII: 1996; **$10-12**

MPC Reissues

1-1701 (462) Superman: 1984; **$30-40**
New larger head parts.

1-1702 (467) Batman: 1984; **$30-40**
New head parts.

Tsukuda Reissues
Produced for Tsukuda by Monogram in Monogram box with Tsukuda logo.

SOT-001-3000 (817) Flying Sub: 1989; **$45-50**
SOT-003-2700 (813) UFO: 1989; **$45-50**

Polar Lights

In 1995 Playing Mantis began reissuing Aurora models made from new steel molds created by copying Aurora's original kit parts. Since then Polar Lights has sometimes used the original tooling owned by Revell-Monogram for reissues. Kits are boxed in facsimile boxes, usually with the "Polar Lights" name replacing "Aurora."

4100 (477) Spider-Man: 2003; **$15-18**
White or red styrene plastic. One kit in twelve is molded in red plastic, visible through oval cutout in box. The original 1/12 scale Aurora model was upscaled to 1/8 to conform to Aurora's original standard figure size.

4101 (421) The Incredible Hulk: 2003; **$15-18**
White or green styrene plastic. One kit in twelve is molded in green, visible through oval cutout in box. The original 1/10 scale Aurora model was upscaled to 1/8.

5001 (805) Addams Family House: 1995-96; **$100-120**
Gray ABS plastic. Released in the fall of 1995 and sold only through FAO Schwarz at $79.

5002 (805) Addams Family House: 1996-03; **$16-20**
Opaque glow plastic. "Frightening Lightning" edition released in the summer of 1996 with glow parts and sold in hobby shops at $13.

5003 (459) Mummy's Chariot: 1996; **$90-100**
Gray ABS plastic. Released early in 1996 and sold only through FAO Schwarz at $85.

5004 (459) Mummy's Chariot: 1996-03; **$15-16**
Glow plastic. Long box. "Frightening Lightning" edition released in the summer of 1996.

5004 (459) Mummy's Chariot: 1997; **$15-16**
Glow plastic. Square box. Sold at K-Mart.

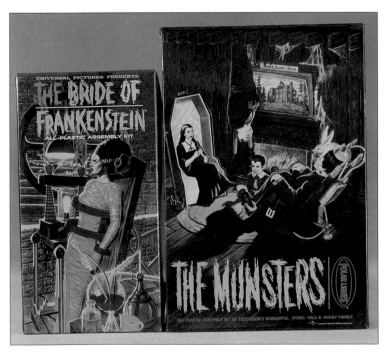

5005 (482) Bride of Frankenstein: 1997-03; **$18-20**
Glow and clear ABS plastic. Second issue in beige styrene plastic.

5006 (465) Frankenstein's Flivver: 1997-03; **$10-12**
Gray ABS plastic. Long box.

5013 (465) Frankenstein's Flivver: 1997; **$10-12**
Gray ABS plastic. Square box.

5013 (804) The Munsters: 1997-03; **$22-28**
Gray ABS plastic. Second issue in beige styrene plastic.

5014 (570) Casper's Undertaker Dragster: 1997-03; **$15-17**
Glow ABS plastic. Second issue beige styrene plastic. Square box.

5015 (458) Wolf Man's Wagon: 1997-2003; **$12-15**
Beige plastic.

5016 (484) King Kong's Thronester: 1998-03; **$14-21**
Beige plastic.

5017 (489) Green Hornet's Black Beauty: 1998-03; **$14-16**
Black styrene plastic. Second issue has chrome parts.

5020 (463) Customizing Monster Kit #1: 1998-03; **$12-14**
Beige plastic. Skull, lizard, *etc.*

5021 (464) Customizing Kit #2: 1998-03; **$12-14**
Beige plastic. Vulture, mad dog, etc.

5026 (466) Dracula's Dragster: 1999-03; **$14-16**
Beige plastic.

5029 (485) The Go Cart: 1999; **$30-35**
Beige and chrome plastic.

5035 (414) James Bond: 1999-03; **$14-16**
Beige plastic.

5036 (415) Odd Job: 1999-03; **$14-16**
Beige plastic.

5030 (418) The Robot from Lost in Space: 1997-03; **$13-19**
Gray ABS plastic. Second issue in beige styrene plastic.

5031 (419) Lost in Space: 1997-03; **$12-17**
Gray ABS plastic. Second issue in beige styrene plastic. Cyclops diorama. Long box without chariot vehicle.

5032 (420) Lost in Space: 1998-03; **$21-26**
Beige plastic. Cyclops diorama. Rectangular box with chariot vehicle.

5081 Planet of the Apes 4-Pack: 2001-03; **$25-28**
Contains all four ape models and cardboard diorama backdrop.

5090 (461) The Bellringer of Notre Dame: 2000-03; **$13-15**
Beige plastic. Produced from a new mold.

5091 (800) La Guillotine: 2000-03; **$13-15**
Beige plastic. Copies of this kit autographed by Tom Lowe were available only to members of the Polar Lights on-line bulletin board members. These came with a certificate of authenticity signed by on-line moderator Lisa Greco and product development manager Dave Metzner.

5092 (483) The Witch: 2000-03; **$15-18**
Beige and clear plastic.

6801 (657) Rodan: 2000-03; **$9-11**
Beige plastic. Aurora logo. Made from new mold.

6802 (658) King Ghidorah: 2000-03; **$9-11**
Beige plastic. Aurora logo. Made from new mold.

6803 (101) Planet of the Apes Cornelius: 2002-03; **$9-11**
Beige styrene plastic. Aurora logo.

6804 (105) Planet of the Apes Dr. Zira: 2002-03; **$9-11**
Beige styrene plastic. Aurora logo.

6805 (102) Planet of the Apes Dr. Zaius: 2002-03; **$9-11**
Beige styrene plastic. Aurora logo.

6806 (103) Planet of the Apes General Ursus: 2002-03; **$9-11**
Beige styrene plastic. Aurora logo.

5093 (818) Dick Tracy: 2000-03; **$14-16**
Beige plastic.

5097 (819) Dick Tracy Space Coupe: 2000-03; **$14-16**
Beige plastic.

5099 (707) Seaview from Voyage to the Bottom of the Sea: 2002-03; **$14-20**
Black or white styrene plastic. One in twelve kits molded in white plastic, visible through oval cut-out in box.

6902 (810) Batcycle: 2003- ; **$13-17**
Dark blue or white plastic, with chrome parts. One kit in twelve is molded in white, visible through oval cutout in box.

6905 (487) Batplane: 2002-03; **$14-18**
Black or white plastic. One kit in twelve is molded in white, visible through oval cutout in box.

6906 (811) Batboat: 2003- ; **$13-17**
Dark blue or white plastic, with chrome parts. One kit in twelve is molded in white, visible through oval cutout in box. The original 1/32 scale Aurora model was upscaled to 1/24.

7501 (426) The Creature: 1999-03; **$16-20**
Dark green styrene plastic. Manufactured from original mold by Revell-Monogram. Aurora logo used with permission of CineModels. "Universal Studios Monsters" logo on box top was not on original Aurora issue. Kits in deeper box were sold exclusively at Toys R Us.

7502 (469) Godzilla: 2000-03; **$15-17**
Lime green plastic. Manufactured from original mold by Revell-Monogram. Aurora logo.

7508 Monsters Frightening 4 Pack: 2000
Contains Frankenstein, Dracula, the Wolf Man, and the Creature. Cardboard display backdrop.

7509 (422) The Forgotten Prisoner: 2001-03; **$17-18**
Gray plastic. Manufactured from original mold.

7503 (427) The Mummy: 1999; **$14-18**
Beige plastic. Aurora logo and original 427-98 marking on box end. Manufactured from original mold by Revell-Monogram. Toys "R" Us exclusive.

7504 (423) Frankenstein: 1999; **$14-18**
Beige styrene plastic. Aurora logo and original 423-98 marking on box end. Universal Movie Monsters on box top. Manufactured from original mold by Revell-Monogram. Toys "R" Us exclusive.

7505 (425) Wolf Man: 1999; **$14-18**
Brown plastic. Aurora logo and original 425-98 marking on box end. Manufactured from original mold by Revell-Monogram. Toys "R" Us exclusive.

7506 (424) Dracula: 1999; **$14-18**
Gray plastic. Aurora logo and original 424-98 marking on box end. Manufactured from original mold by Revell-Monogram. Toys "R" Us exclusive.

7507 (468) King Kong: 2000-03; **$16-18**
Gray plastic. Manufactured from original mold by Revell-Monogram. Aurora logo. Small palm tree parts are missing.

7512 (816) Land of the Giants snake scene: 2002-03; **$14-18**
Green or white plastic. One kit in twelve is molded in white, visible through oval cutout in box.

7513 (830) Land of the Giants Spaceship: 2002-03; **$18-20**
Orange or white and lime green, clear plastic. One kit in twelve is molded in white, visible through oval cutout in box.

Addar

This company operated from 1972-76 under the direction of ex-Aurora president Abe Shikes and many of the former leaders of Aurora.

Aircraft

901(292, 287) F4D Skyray and **F3D Skynight**: 1976; **$20-25**
Ex-Comet. Issued by Aurora as **292** and **287**.

902 (289) F-100 Super Sabre and **F-94 Starfire**: 1976; **$20-25**
Ex-Comet. F-100 issued by Aurora as **289**. F-94 issued by Aurora Canada as **2495**.

903 (293, 299) F9F-6 Cougar and **F-84 Thunderstreak**: 1976; **$20-25**
Ex-Comet. Issued by Aurora as **293** and **299**.

904 (296, 297) KC-135 Supertanker and **B-58 Hustler**: 1976; **$20-25**
Ex-Comet. Issued by Aurora as **296** and **297**.

Ships

The ship-in-a-bottle kits were originally issued by Gowland in 1953. Patterns for the ships were sculpted by Derek Brand, later head of Aurora's slot car division. The leased molds disappeared with Addar's demise. Prices are **$20-30**.

201 Flying Cloud
202 Santa Maria
203 Revenue Cutter
204 Bon Homme Richard
205 Constitution
206 Mayflower
207 Charles W. Morgan
208 Savannah

Super Scenes

220 Prehistoric Dinosaurs: 1975; **$20-30**
Scene in a Gowland bottle.

225 World War I Dogfight: 1975; **$25-30**
Fokker Triplane and Sopwith Triplane in a Gowland bottle.

226 Rendezvous In Space: 1975; 1/60; **$25-30**
Apollo-Soyuz space capsules.

227 Spirit in a Bottle: 1975; **$40-50**
Scene in a Gowland bottle.

215 Tree House: 1975; **$30-35**
Scene in a Gowland bottle. Box art by Schaare.

216 Cornfield Roundup: 1975; **$30-35**
Scene in a Gowland bottle. Box art by Schaare.

217 Jail Wagon: 1975; **$30-35**
Scene in a Gowland bottle. Box art by Schaare.

Figure Kits

Based on the *Planet of the Apes* movies. All are tan plastic. An eighth figure, Galen, was announced but not issued. Three *Planet of the Apes* "scenes" were put in Gowland bottles.

152 Evel Knievel: 1975-76, 1/12; **$80-100**
Evel and his motorcycle. Box art by Schaare.

153 Evel Knievel's Wheelie: 1975-76, 1/12; **$80-100**
Box art by Schaare.

154 Evel Knievel's Sky Cycle: 1975-76, 1/24; **$80-100**
Diorama of Snake River Canyon. Box art by Schaare.

101 Cornelius: 1973; **$50-60**
Box art by Schaare. Reproduced by Polar Lights.

102 Dr. Zaius: 1973; **$50-60**
Box art by Schaare. Reproduced by Polar Lights.

103 General Ursus: 1973; **$50-60**
Box art by Schaare. Reproduced by Polar Lights.

104 General Aldo: 1973; **$50-60**
Box art by Schaare.

105 Dr. Zira: 1974; **$50-60**
Box art by Schaare. Reproduced by Polar Lights.

106 Caesar: 1974; **$60-70**
Box art by Schaare.

107 Stallion & Soldier: 1974, 1/16; **$200-240**
Box art by Schaare.

231 Jaws: 1975; **$40-50**
Scuba diver and shark in a Gowland bottle. Not to be confused with **270**.

250 Blue Jays: 1975; **$15-20**
Male and female in a bottle.

251 Red Cardinals: 1975; **$15-20**
Male and female in a bottle.

253 Outlaw Mustang: 1975, 1/16; **$30-35**
Same horse as in **107 Soldier on Stallion**.

270 Jaws: 1975, 1/20; **$100-120**
Not to be confused with **231**. Shark's head breaking surface of the water with a cardboard backdrop of sinking fishing boat.

Toy Cars of Japan & Hong Kong. Andrew G. Ralston. This book presents a unique selection of the most rare and collectible toy cars made in Japan and Hong Kong in the 1950s and 1960s. Three particular groups of toys are covered: the magnificent large-scale tinplate cars made for the American market; the smaller but equally sought-after Japanese diecasts by Model Pet, Micro Pet, and Cherryca Phenix; and plastic friction-drive cars made in Hong Kong. Some of the toys pictured are so scarce that they are believed to be among a mere handful that survive in mint condition with their original boxes. This book provides many fascinating new insights to the history of the companies that made and distributed the toys, using long-forgotten trade journals, rare catalogs, and interview with people who worked in the toy business at that time.

Size: 8 1/2" x 11"	340 color photos	160pp.
ISBN: 0-7643-1196-4	hard cover	$39.95

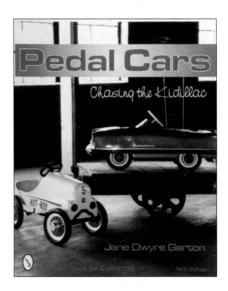

Pedal Cars: Chasing the Kidillac. Jane Dwyre Garton. A fantastic presentation of history and culture, of designers and automobiles. The author has explored family history and far beyond to explain how pedal cars were made, and why they ultimately lost favor in the marketplace. Workers who actually made the cars provide information that counts for collectors. Charming oral histories from the men and women who made hubcaps and welded little bodies come to life in photos and illustrations from the factories to the showroom floor. See hundreds of cars that are lovingly collected in all stages of repair, from rusty frames and rotting wheels to gleaming chrome and custom paint. It is the first compilation of photos and facts put together to tell a story about where the toys were made and how they continue to give joy to collectors, in addition to up-to-date pricing information.

Size: 8 1/2" x 11"	550 photos	208pp.
ISBN: 0-7643-0836-X	hard cover	$49.95